EXCITING ICT IN
MATHS

Alison Clark-Jeavons

Published by Network Educational Press Ltd
PO Box 635
Stafford
ST16 1BF

First published 2005
© Alison Clark-Jeavons 2005

ISBN 185539 191 0

Managing editor: Mary Pardoe
Cover design: Neil Hawkins, NEP
Layout: Neil Hawkins, NEP
Illustrations: Katherine Baxter

Printed in Great Britain by MPG Books Ltd, Bodmin, Cornwall

Contents

Foreword

I have had the great honour and pleasure to travel a lot of the world for over a decade talking, advising and consulting about ICT and learning. It is a personal passion and, being married to a teacher, an issue that concerns me both professionally and personally. When I started I met many enthusiasts who shared a dream that we could use ICT to transform the experience of education and deliver a global aspiration of education for every citizen of the world. Against these lofty ambitions the perennial constraints of budgets, political will and professional inertia were easily visible. There were many fears expressed to me. Most importantly, there was a concern that, somehow, ICT would be used as an excuse to sack teachers and close schools and pump learning into kids' heads through impersonal technology.

Being an optimist, my experience over the last decade has kept those lofty aspirations alive. It has also made the fundamental truths about both education and learning clear to me. First, learning is at its heart a social and a socializing experience. ICTs are very powerful tools, but smart technologies need smart people, they don't replace them. In a world where technology is increasingly pervasive, teachers become more not less important.

Second, as the world becomes increasingly connected, as technology and science develop at an ever-increasing pace, the economic and social future of any country is increasingly tied to its commitment to education and training, not just for the elite but for every citizen and community.

Third, the goal is not just raising standards but changing culture. I describe this using the analogy of the driving test, a rite of passage for many young people. The emerging global information society requires us to create a new generation who, when they leave school, put on their L-plates and think 'I am a learner', rather than take them off and say 'I have passed'.

We can only make this happen on the scale needed if we value and invest in our teachers as lifelong learners themselves, not just in their 'subject skills'. To do this we need to marry the big picture of a transformed experience of learning to ICT practice, but also to new theories of learning such as learning styles or multiple intelligences. For teachers to be seen as learners themselves we need to build bridges between different areas of research – in education, learning theories, ICT and management, to name but a few.

Reading the first few titles in this series, it is wonderful to see words like creativity, personalization and exciting being based on actual evidence, not just lofty aspiration. The rate of change of technology in the next decade will at least match the progress in the last. The materials available to enrich good teaching and learning practice will grow exponentially. None of this will have the profound change that many aspire to if we cannot build the bridge between theory and what happens in individual lessons, be they in art, maths, music, history, modern languages or any other area of the curriculum.

The notion of ICT as a tool across the curriculum was greeted sceptically a decade ago. Many professionals told me that ICT may be important in maths or science, but irrelevant in the arts and humanities. My own experience is that the most exciting innovations have actually been in arts and humanities, while the notion of maths as a visual discipline seemed alien a few years ago. It has not been ICT but innovative teachers, researchers and indeed publishers who have pushed the art of the possible.

In a lot of my work, I have encouraged the notion that we should see the era we live in as a New Renaissance, rather than a new Industrial Revolution. While the industrial revolutions were about simplification and analysis, the era we live in is about synthesis and connection. We need our learners to embrace both depth and breadth to meet their needs to learn for life and living.

To the authors of this series, I offer my congratulations and sincere thanks. In bringing together the evidence of what works, the digital resources available and the new theories of learning, along with the new capabilities of ICT, they bring the focus onto the most important element of the transformation of learning, which to me is the learning needs of the teaching profession.

To the readers of this series, I make what I believe is my boldest claim. This is the greatest time in human history to be a teacher. Our societies and economies demand education like never before. Our increasing knowledge of how we learn and how the brain works, together with the availability of powerful ICT tools, make this a time when the creativity, professionalism and aspirations for a learning society are at a premium. Teaching is a noble profession. It is after all the profession that creates all the others.

There are many things that we do not yet know, so much to learn. That is what makes this so exciting. I and my colleagues at Microsoft can build the tools, but we believe that it is putting those tools in the hands of innovative, skilled and inspirational teachers that creates the real value.

I hope that after reading any of the books in this series you will feel the excitement that will make learning come alive both for you and the children you teach.

Best wishes

Chris Yapp
Head of Public Sector Innovation
Microsoft Ltd

Author's acknowledgements

My first thanks need to go to my Mum and Dad, whose purchase of a Sinclair ZX81 probably began my journey with technology in mathematics – their later purchase of an Amstrad 286 some 15 years later helped me through my PGCE course and trying to communicate mathematically using WordPerfect!

I would also like to express my gratitude to Dianne Smith, headteacher of Admiral Lord Nelson School, Portsmouth, and Dame Sheila Wallis, former headteacher of Davison High School for Girls, Worthing. As inspiring leaders, with clear educational visions, they both gave me sustained opportunities for ongoing professional development, which I believe to be an entitlement for all teachers throughout their careers. The staff at The Mathematics Centre, University College Chichester, in particular Professor Afzal Ahmed and Honor Williams and, more recently, Debbie Yates, have been central to this process and I appreciate their ongoing support and encouragement.

The two people most instrumental in opening my eyes to the potential of technology in mathematics education were Adrian Oldknow and Warwick Evans. Adrian continues to be an inspirational colleague and much of the content of this book has arisen from our most inquisitive moments when working together. All of us who knew and worked with Warwick are only too painfully aware of the gap he has left behind as a mathematician, teacher, musician and friend, and can only ponder what he might have created with the current mathematics toolkit.

Finally, to all the teachers whom I have the privilege of supporting on their individual professional development journeys, a big thank you for your honesty and openness which enables all of us to learn more about teaching and learning mathematics, with or without technology.

Introduction

Most days I run by the sea. If the tide is out, I run on the beach and the most striking feature of this is that the landscape is never the same two days in a row. Even so, there are some things that are constant; the breakwaters, the lighthouse and the position of the buoys. But it is the variables that mean it could never be boring! The waterline changes, the rocks and pebbles get moved about, the seaweed comes and goes and sometimes debris is washed up. And then there's the sky – clouds changing shapes, light coming and going and the sun changing its position in the sky.

The question, what is changing and what stays the same is the only question that I need to ask to begin a journey of mathematical discovery. In my mind I have often transported myself back in history and become a teacher of mathematics on a beach in a time before schools had been invented. The thought of a group of pupils, with nothing more than their natural inquisitiveness, questioning the variables and constants in their surroundings, as the early mathematicians did, is an exciting prospect.

But that is a long way from our current situation with its formalized learning environments, extensive assessment systems, league tables, inspections and accountability! How do we stimulate and maintain pupils' natural curiosity in such a way that they are not switched off mathematics by the time they reach Key Stage 4?

It is my view, and that of many others, that ICT offers a route to this aim.

This book aims to provide a stimulus for you to review *how* you can use ICT to deepen pupils' understanding of mathematics. By considering some of the processes through which we learn mathematics, such as visualizing, making connections and observing similarities and differences, different ICT tools are introduced and described. Case studies from Key Stages 2, 3 and 4 classrooms, as well as experiences of teachers in professional development activities, exemplify how some of the ideals can be achieved in practice.

This book is intended to widen your knowledge about both the range of ICT tools available and the different ways in which they can be used. It is neither possible nor appropriate to look at every available resource. For example, there is no reference to the use of spreadsheets, not because I don't believe that they can be used effectively in the mathematics classroom, but because I have preferred to focus on other tools designed specifically for mathematics.

You cannot reflect upon any aspect of the teaching and learning of mathematics without actually *doing* some! That is the equivalent of trying to teach someone the technicalities of playing football in a classroom, without ever kicking a ball.

The accompanying CD-ROM contains files for you and your pupils to explore together including trial versions of some of the software featured. There is no doubt that real learning involves taking a risk. It would be an easy ride just to read the book and be impressed by the amazing things that ICT can make happen in mathematics. But, if you would like to begin to use ICT in your mathematics classroom, you will need to invest some time and effort.

So, find yourself a computer, put in the CD-ROM and start reading!

Chapter 1

ICT and mathematics – evaluating effectiveness

In this section you will:

■ consider some criteria for evaluating the potential of ICT to support the learning of mathematics;

■ find out about some recent research into learning mathematics using ICT;

■ preview what is coming!

The idea of 'talking mathematics' to a computer can be generalized to a view of learning mathematics in 'Mathland'; that is to say, in a context which is to learning mathematics what living in France is to learning French.

Seymour Papert

Some mathematics educators and mathematicians have been enthusing, for more than 20 years, about the way ICT will revolutionize the teaching and learning of mathematics. In 1980, Seymour Papert published his book *Mindstorms: Children, computers and powerful ideas* in which he describes how computers enable pupils to enter his 'Mathland' called LOGO.

The use of LOGO is still referred to in many national curricula. In England and Wales the Key Stage 3 *Framework for Teaching Mathematics: Years 7, 8 and 9* recommends LOGO within exemplar activities for the exploration of angle properties of polygons.

This prompts the questions:

➡ What does LOGO offer that has enabled it to 'stay the distance'?

➡ What is it that pupils experience in this microworld?

➡ Are any of these features exemplified in other microworlds?

Papert describes a microworld as an 'incubator of knowledge' in which pupils experience a set of constraints that are consistent with some mathematical constraints.

Using ICT within mathematics is not a new idea. We have had access to television for over 40 years, and the overhead projector is still an important resource for many teachers. Back in the 1980s it was not unusual for primary and secondary schools to have BBC Acorn computers. Mathematics departments used these computers with a host of software titles that supported mathematics teaching and learning. Much of this original software had been written and developed by teachers and, although it may have been limited by the lack of a mouse or sophisticated graphics, it provided stimulating mathematical environments.

In the mid-1990s, two factors drastically reduced access to ICT for mathematics:

➡ information technology was designated a national curriculum subject in 1995, which meant that computer suites were needed for timetabled IT lessons;

➡ many schools switched to PCs with Microsoft®-based software which could not run the existing mathematics software.

It is only now that we find ourselves in a Renaissance period in which the access to technology has re-emerged *and* there is a range of new software available. The difference now is that teachers are expected to be far more discerning about how they choose to incorporate ICT into their teaching. Questions such as, 'How is the ICT enhancing the learning' and 'How is the ICT being used across the curriculum?' are on every Ofsted inspector's lips!

The internet, and a fast connection to it, is undoubtedly a route to an extensive range of information, free software and resources. For many teachers, it is the first consideration when looking for resources to support learning.

Take an aspect of learning about fractions at Key Stages 2 or 3. What use could the internet be?

WEBSITE

Typing 'Fractions' into a search in Google revealed 1,190,000 results in 0.29 seconds. On the day that I searched, the first link was to http://www.visualfractions.com/ described as 'an online tutorial that offers instruction and practice in identifying, renaming, and operating on fractions'. I chose an activity called 'Identify with lines'.

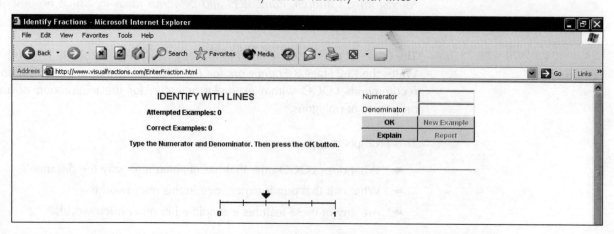

This appeared to be a straightforward 'drill and skill' question in which I was expected to input a numerator and denominator and submit my answer.
So, how clever is the program sitting behind this web page?
I tested it out…

➡ What happens if I input the correct answer, Numerator = 2 and Denominator = 5? (The response is '2/5 is correct'!)

➡ What happens if I input an alternative correct answer, such as the equivalent fraction with Numerator = 4 and Denominator = 10? (The response is '4/10 is correct'!)

➡ What happens if I input a wrong answer, Numerator = 2 and Denominator = 3, which would indicate a common misconception? (The response is '2/3 is too large' and I can have another attempt.)

➡ What about a completely wrong answer such as Numerator = 1 and Denominator = 5? (The response is '1/5 is too small' and I can have another attempt.)

The 'Explain' option gave the correct answer and explained, in text, how the numerator and denominator had been worked out. There was no audio option and no additional graphics.

So how might I plan to use this free web-based piece of software in my teaching?

➡ For which pupils is it appropriate?
➡ How would they access the activity?
➡ For how long should they use the software?
➡ What, if anything, would they record?
➡ What support resources might they need?
➡ What feedback would I receive about pupils' progress and any misconceptions they might have?
➡ Do I have any control over the fractions that are generated?
➡ What ICT skills might the pupils (or I) require in order to use the software?

I stayed with this activity a little longer. After about five minutes, and 39 examples later, I realized that the examples were being randomly generated, the denominators did not go above 12 and, as I have a good understanding of 'recognizing fractions on a simple zero to one number line', I was extremely bored!

I was doing very little thinking, certainly had no reason to discuss my ideas with anyone else, and I was not using any creative skills.

So you could:

➡ give this activity to pupils who already have a good understanding of the objective for, say, ten minutes at a time, to assess their skills;

or perhaps,

➡ use it as a review activity during a plenary, when each example could be displayed through a data projector and the pupils asked to write and display their responses on individual whiteboards.

But I certainly wouldn't plan to use valuable ICT access to take a group of pupils to the ICT suite for a lesson on this particular activity!

On the same site I found an activity called 'Find Grampy', which, although it addresses the same learning objective, would seem to offer a much more interesting task.

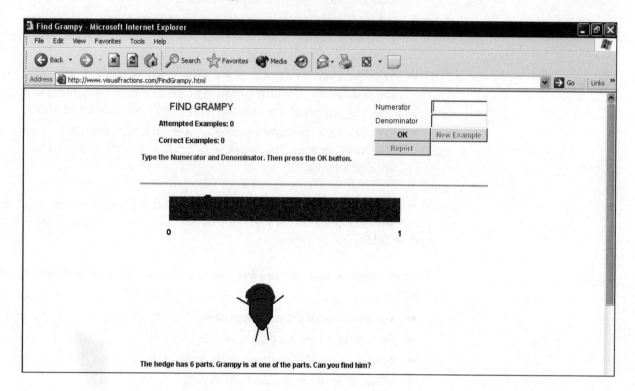

To start with there are no markings on the hedge; I could envisage pupils working in pairs, discussing how they might divide the hedge and estimating the position. They may fold paper strips of the same length as the hedge to check their estimates. All in all it seems a better choice of activity.

So how do we decide if the ICT is enhancing the mathematical learning?
If we think back to Seymour Papert's words, this activity does not seem to be an 'incubator of knowledge', although most teachers could see some advantages for pupils in using it.

And that was just the first of the 1,190,000 links from the Google search – what about the other 1,189,999?

Some criteria to consider

Clearly, it would be useful to have a set of criteria against which to make some judgements about when and how to use ICT to support learning.

Many mathematics educators and software developers say that they are trying to design microworlds that provide insight into how we manipulate different mathematical ideas and concepts.

Paul Goldenberg writes, in *Issues in Mathematics Education*:
'It is the problems that are posed, not the technology with which they are attacked, that makes all the difference. With computers, as with pencils, some problems are great and some are a waste.'

Paul Goldenberg has developed six principles for thinking about technology in mathematics classrooms. These are:

The *genre* principle
Consider the different roles for technologies. Think clearly about the classroom goals and choose technologies expressly to further these goals.

The *purpose* principle
Consider the purpose of the technology, for example to speed up calculations and thus facilitate exploration.

The *answer v analysis* principle
Consider if the technology is being used to solve a problem, or find a result, or is it being used to help pupils think about a problem.

The *who-does-the-thinking* principle
Consider whether the technology is replacing a capacity that the pupil may need to develop, or developing the pupil's capacity to think.

The *change-content-carefully* principle
Consider thoughtfully what is or is not obsolete content. Don't just think about what the technology can do, but carefully analyse what students need to be able to do, and especially how they need to be able to reason.

The *fluent tool user* principle
Consider teaching pupils to use a few good tools well enough to use them knowledgeably, intelligently, mathematically, confidently and appropriately in solving otherwise difficult problems.

The set of principles above is fairly challenging. To make it more meaningful we will consider a specific learning objective and develop an ICT-based lesson that seems to meet the principles.

All the principles assume that you already have a clear idea of the overall purpose of the lesson – if you are not clear about this, you cannot possibly begin to choose the most appropriate resources to support the learning!

Let's take a learning progression for Shape, space and measures: Transformations from Year 4 to Year 7:

Sketch the reflection of a simple shape in a mirror line parallel to one side (all sides parallel or perpendicular to the mirror line).

(Year 4)

Recognize reflection symmetry in regular polygons: for example, know that a square has four axes of symmetry and an equilateral triangle has three.

Complete symmetrical patterns with two lines of symmetry at right angles (using squared paper or pegboard).

Recognize where a shape will be after reflection in a mirror line parallel to one side (sides not all parallel or perpendicular to the mirror line).

(Year 5)

Recognize where a shape will be after reflection:

■   in a mirror line touching the shape at a point (sides of shape not necessarily parallel or perpendicular to the mirror line);
■   in two mirror lines at right angles (sides of shape all parallel or perpendicular to the mirror line).

Recognize where a shape will be after two translations.

(Year 6)

Understand and use the language and notation associated with reflections, translations and rotations.

Recognize and visualize the transformation and symmetry of a 2-D shape:

■   reflection in given mirror lines, and line symmetry;
■   rotation about a given point, and rotation symmetry;
■   translation;

explore these transformations and symmetries using ICT.

(Year 7)

If these are the outcomes that we want, how can we design a learning environment for pupils that will help them *understand* the concept of reflection?

Immediately, as teachers, we may begin to think about the many experiences of reflection that pupils have had in their lives. What obvious mathematical and scientific tools can we use in the classroom that will relate these experiences to our learning goals? (The nature and role of tools to support learning mathematics is discussed in more detail in Chapter 3.)

In the classroom we use mirrors, folded paper, tracing paper, images and objects to try to mathematicize pupils' real-life experiences and introduce the mathematical vocabulary that our curriculum requires them to master.

How might ICT enhance these experiences? And in a way that is consistent with Paul Goldenberg's six principles?

Taking his last criteria as a starting point, it is sensible to choose an ICT tool that will allow all pupils to explore, during Key Stages 2 and 3, the selected objectives, and perhaps others too.

One such ICT tool is dynamic geometry software which was developed in the late 1980s and early 1990s simultaneously in the USA (The Geometer's

Sketchpad), France (Cabri Geometry) and Israel (The Geometric Supposer). Such software is explicitly recommended in the Key Stage 3 Framework. However, few Key Stage 2 teachers have come across it.

Dynamic geometry software offers an interactive environment in which pupils can explore a wide range of mathematical ideas, including symmetry, graphs, shapes, area and perimeter. It is possible for them to work within a pre-written task, or design their own solutions to problems they have been set.

CD-ROM

The Becta website gives some background information on dynamic geometry software and the CD-ROM accompanying this book contains a 30-day trial version of The Geometer's Sketchpad and the software files.

Having chosen a tool, let's think about what the activities could look like from Year 4 through to Year 7.

CD-ROM

If you can, load the software onto your computer and open the file *Transformation Y4 to Y7.gsp* so that you can play as you read!

Year 4 'Mrs Reflect'

In Year 4, the teaching builds on pupils' everyday experiences of reflection.

A simple profile of a face which can be reflected in the mirror line by the software provides an exploratory environment for pupils to ask 'what if …?' questions:

What if we alter Mrs Reflect's face and make her unhappy?

What if we move her away from the mirror?

…and what if we move the mirror line?

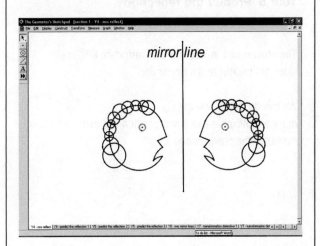

Year 4 Predict the reflection

In this task, pupils drag black tiles from a pile to design their own shape or pattern on one side of the mirror line. They then predict where the reflection will be, and drag grey tiles to show their prediction.

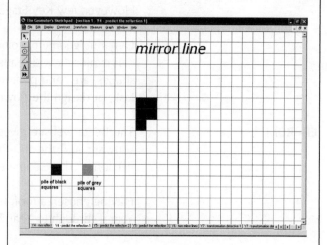

Pupils then check their prediction by using the software to reflect their original shape. If necessary they can reposition any grey tiles at this stage.

What if we move the mirror line?

In both examples, it is important that the mirror line can be moved to a different position or orientation on the page.

In this way pupils experience the effects of moving the mirror line in relation to the original shape, even though this may not be a focus for the teaching in Year 4.

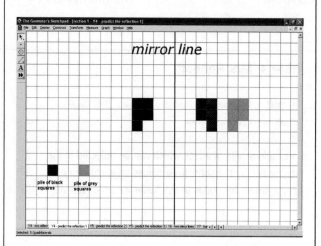

Year 5 'Predict the reflection'

The progression in Year 5 lets us introduce the idea of sides not parallel to the mirror line.

We can do this in two ways…
move the mirror line to a new position and carry out a task similar to the previous one…

…or…

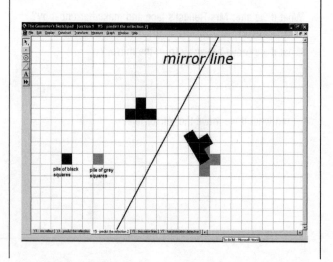

...include polygons that do not necessarily have any right angles.

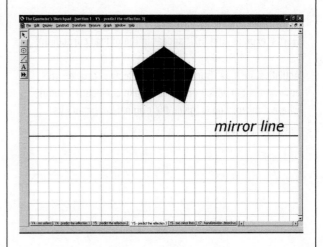

Year 6 'Double reflect'

In Year 6, pupils explore what happens if a shape touches the mirror line – although it is likely that their curiosity will have led them to explore this in previous activities.

In the activity 'Double reflect' the 'snap to grid' facility supports pupils as they design a polygon and reflect it twice.

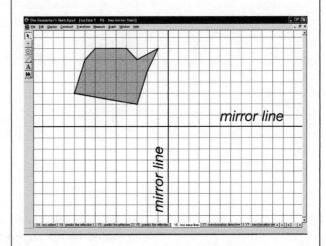

Pupils can then be encouraged to explore what happens if they distort their original shape.

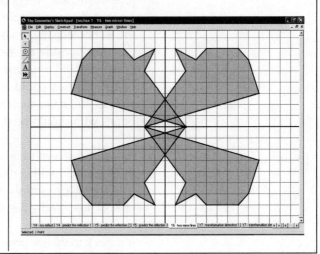

Year 7 'Transformation detective'

Pupils are expected to have even more autonomy – the words 'explore' and 'ICT' appear in the official curriculum!

Activities such as 'Transformation detective' provide an environment in which pupils can make conjectures about where a mirror line is, given the original image and its reflection.

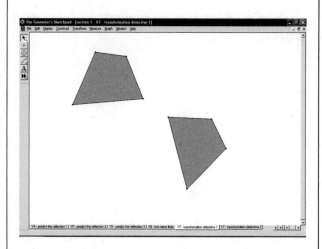

Or pupils might explore more challenging situations; given an original image and its rotation, what was the angle of rotation and where was the centre of rotation?

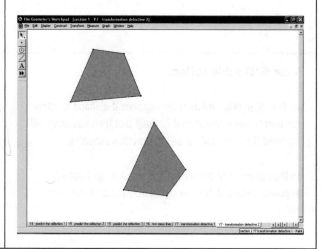

So let us briefly reflect on Paul Goldenberg's six principles. In this series of activities the ICT supports pupils in exploring transformations using a range of approaches. The technology gives pupils accurate feedback without passing judgement, and the pupils are using the ICT in a way that allows them to see many different examples and begin to hypothesize for themselves. The use of the technology might also suggest that younger pupils can begin to access what could be considered 'extension' ideas in an informal setting, or the use of ICT could demand a review of the timings within the curriculum. Of course, all of these activities could be 'demonstrated' to pupils in a whole-class teaching setting. But, if we are sticking close to Goldenberg's criteria, it is very much the pupils who should be driving the software!

As yet, we haven't considered the motivational aspects of ICT. We know that most pupils would certainly prefer to learn at a computer rather than work from a textbook or worksheet. And if the computer collects their responses and checks them and gives the teacher feedback, surely that must be a good thing?

do we?

It is back to the six principles. If the purpose of the lesson is to assess learning, then an ICT-based 'skill and drill' lesson might seem to be appropriate.

However, a rich problem-solving task in which pupils use an ICT tool to formulate their own response may provide far richer assessment data. The act of using the tool may also support pupils' mathematical understanding.

Chapter 2 looks at ICT tools that support assessment opportunities in more detail.

What does the research say?

The emerging knowledge base in relation to the use of ICT in mathematics has some clear things to say...

In every case except one the study found evidence of a positive relationship between ICT use and achievement.

ImpaCT2 The Impact of Information and Communication Technologies on Pupil Learning and Attainment. ICT in Schools Research and Evaluation Series – No.7

WEBSITE

http://www.becta.org.uk/page_documents/research/ImpaCT2_strand1_report.pdf

There is a strong relationship between the ways in which ICT has been used and pupils' attainment. This suggests that the crucial component in the appropriate selection and use of ICT within education is the teacher and his or her pedagogical approaches. Specific uses of ICT have a positive effect on pupils' learning where the use is closely related to learning objectives.

It is not what ICT you use but how you use it that determines improvements in learning.

ICT and Attainment DfES ICT in Schools Research and Evaluation Series – No.17

WEBSITE

http://www.becta.org.uk/page_documents/research/ict_attainment_summary.pdf

'Teachers' professional development in technology and the use of computers to teach higher-order thinking skills were both positively related to academic achievement in mathematics.'
Harold Wenglinsky

Does It Compute? The Relationship Between Educational Technology and Student Achievement in Mathematics (1998).

WEBSITE

http://www.monet.k12.ca.us/Challenge/pdf_files/ets_math.pdf

If you want to find out more about current UK research into using ICT to support learning, the following organizations have research areas on their websites with hyperlinks to a variety of recent findings:

WEBSITE

Becta (http://www.becta.org.uk/);
NESTA Futurelab (http://www.nestafuturelab.org);
The REVIEW Project (http://www.thereviewproject.org/);

WEBSITE

Interactive Education (http://www.interactiveeducation.ac.uk/);
London Knowledge Lab (http://www.lonklab.ac.uk/index.html).

Educational journals that also look at issues surrounding mathematical learning using ICT include:

Micromath (Association of Teachers of Mathematics);

The International Journal of Computers for Mathematical Learning (Kluwer);

Educational Studies in Mathematics (Kluwer).

Evaluating 'effective' use of ICT for the learning of mathematics

What follows is an approach that has been designed to help teachers decide whether the use of ICT has been effective in supporting learning. It has been used successfully with teachers on professional development courses, providing a tangible way of thinking about the different aspects of mathematical learning that have occurred in a lesson.

This is offered as a practical task. It was developed from the booklet *Mathematics from 5–16*, which was published in 1985 as part of the HMSO Curriculum Matters series. It was one of the most useful publications given as curriculum guidance to schools. Appendix 1 of the booklet gives the following list of general objectives for a mathematics curriculum:

Facts	Terms Notation Conventions Results
Skills	Performing basic operations Sensible use of a calculator Simple practical skills in mathematics Ability to communicate mathematics The use of microcomputers in mathematical activities
Conceptual structures	Understanding basic concepts The relationship between concepts Selecting appropriate data Using mathematics in context Interpreting results
General strategies	Ability to estimate Ability to approximate Trial and error methods Simplifying difficult tasks Looking for patterns Reasoning Making and testing hypotheses Proving and disproving

Personal qualities	Good work habits: ■ imaginative, creative, flexible; ■ systematic; ■ independent in thought and action; ■ co-operative; ■ persistent. Positive attitudes to mathematics: ■ fascination with the subject; ■ interest and motivation; ■ pleasure and enjoyment from mathematical activities; ■ appreciation of the purpose, power and relevance of mathematics; ■ satisfaction derived from a sense of achievement; ■ confidence in an ability to do mathematics at an appropriate level.

It is quite significant that the section 'personal qualities' did not appear in subsequent national curriculum documents, perhaps because the qualities were perceived as difficult to level or examine!

In order to complete the task, you will need to focus on a particular mathematics lesson in which you have used ICT, and in which you think the ICT made a difference. Try to concentrate on the mathematical learning that you think was enhanced, although it is very likely that the ICT had a positive effect on pupils' motivation and engagement too.

You will need a sheet of A3 or A2 paper and some glue.

Photocopy the resource in Appendix 1 (pages 141–143) onto card and cut out the statements to produce a set of sort cards.

In the centre of your paper, write the mathematical aims of the lesson that you are evaluating and a brief description of what the pupils *actually did*. Be careful, write what they did, *not what you hoped that they would do*! (We'll come back to that later.)

Your evidence for what the pupils did will probably draw from the following list:

➡ your observations of the pupils at work;
➡ overhearing pupils' own conversations and the mathematical vocabulary they chose to use;
➡ your conversations with pupils;
➡ pupils' on-screen responses to the task;
➡ pupils' written responses to the task, if appropriate;
➡ the questions pupils asked you or each other;
➡ pupils' responses to questioning by you or each other;
➡ pupils' attitudes to the task;
➡ the way the pupils worked at the task.

Try to think about the mathematical facts, skills, conceptual ideas and general strategies that the pupils *actually used*.

Now take the set of sort cards and familiarize yourself with all the statements. Focusing on one card at a time, decide whether the pupils achieved the statement or not. If it was achieved, record on the back of the card what evidence you have for your claim. Take time over this – it is often easy to think that pupils have achieved something, but when you are forced to cite your evidence as to how it was achieved, you become less confident in your claim!

You will end up with two sets of cards. For now, put the discarded cards to one side.

On the large sheet of paper begin to arrange the cards for which you have concrete evidence in a way that tells the story of your lesson. This arrangement will be unique to you. It should be a map of the different aspects of the lesson and you are likely to find that the cards link in some way.

For example, you may decide that within the lesson you observed that pupils were independently looking for patterns and, on obtaining results, showed a fascination for the subject.

If you are able to decide on a final position for the cards on the page, stick them on and annotate the page to include your thoughts about how the cards connect together.

So far, we have not considered where the ICT has particularly enhanced the pupils' experiences, so the big question to ask is:

> **Which of the statements on the page was achieved because ICT was the predominant mathematical tool chosen for the task?**

Colour code or mark the cards showing the statements that you select.

You have now deeply evaluated the mathematical learning that took place in your lesson and have a rationale to decide how the ICT enhanced this learning. The final stage of any evaluation process is to consider what you would do differently next time.

This is where the discarded cards are useful again. Go back through this set of cards and select those that state things that you would have liked to have seen the pupils achieve in the lesson. Think about how you would change the lesson to accommodate these revised aims.

Essentially, this task has helped you to 'find out where you are at'! You have established your own prior learning within the context of planning to use ICT to support mathematics teaching and learning.

This book aims to engage you actively in developing your practice by extending the range of ICT-based approaches that you use in your classroom, while firmly keeping an eye on how its use enhances mathematical learning. Appendix 2 (page 144) is a blank checklist included for your use. Make some photocopies to keep beside you as you plan your lessons. Aim to include as many of the objectives as you consider appropriate to the mathematical topic and your pupils.

A word about creativity

An emerging feature of some of the most exciting uses of ICT in mathematics is that they facilitate creativity.

Here is a simple example.

Chapter 1 opened with a reference to LOGO and discussed the way in which it has been embraced as a resource to support pupils' understanding of the concept and applications of exterior angles of polygons.

A few years ago, Adrian Oldknow and a team of authors based at University College Chichester, were authoring the content for the Research Machines (RM) MathsAlive Key Stage 3 scheme of work. Adrian and I were faced with the task of developing some new ICT-based approaches to the Year 9 learning objective, 'Explain how to find, calculate and use: – the sums of the interior and exterior angles of quadrilaterals, pentagons and hexagons; – the interior and exterior angles of regular polygons'.

We had already included a LOGO-style activity and decided to consider how we might use The Geometer's Sketchpad to create an alternative, but complementary approach.

This is what we came up with:

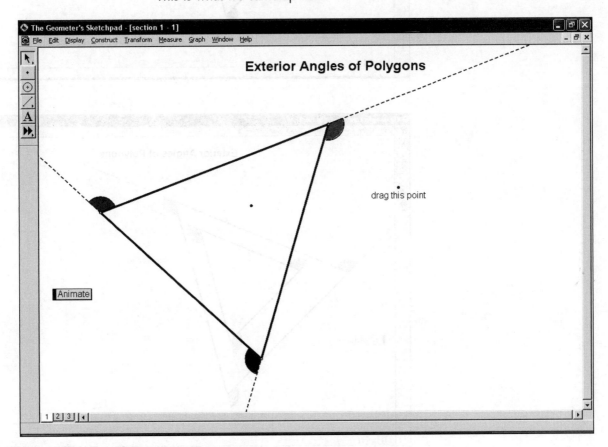

The image above shows the opening screen.

Look carefully at it – what do you think is going to happen?

The title gives you a clue, and also the 'Animate' and 'drag this point' text.

CD–ROM

If you have access to the CD-ROM, load The Geometer's Sketchpad file *Exterior angles of polygons.gsp* and click 'Animate' or 'drag the point'.

Is that what you expected?

For those who do not have access to the CD-ROM, look at the sequence of images below.

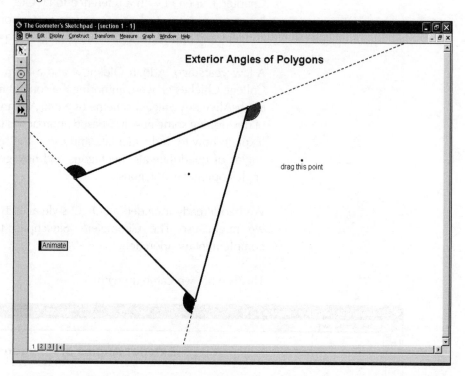

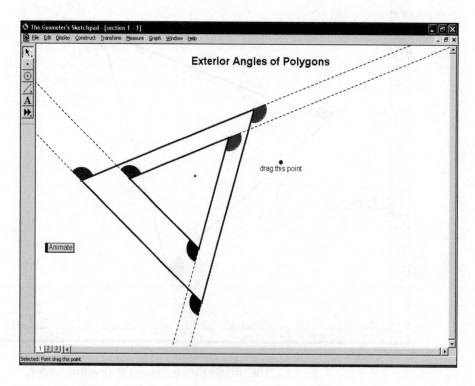

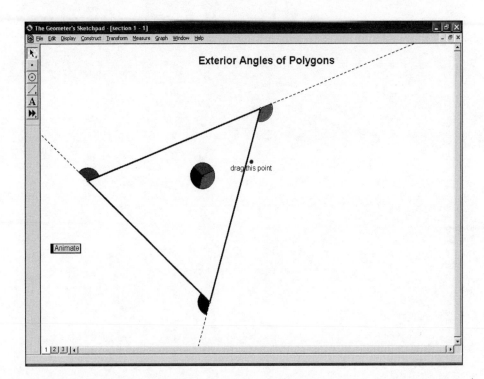

I have used this dynamic image with groups of teachers. Most admit that they have never imagined that the proof for the exterior angles of any polygon could be shown so simply, by considering it as an enlargement! Perhaps this is because we have not had that imagery presented to us in the static pages of our textbooks, or maybe because we were taught transformations in a different context.

The next step is to try dragging the various points to different positions, and experimenting, until you convince yourself visually that the exterior angles of a triangle will always sum to 360 degrees. And then what about quadrilaterals, pentagons…?

Many teachers ask 'How did you create this image', and are keen to have a go at re-creating the image.

This suggests that it is not enough just to watch someone else's creation. Often the real learning experience happens when you try to re-create something for yourself. This is another key feature of an effective ICT tool – it lets you take control! We will return to the idea of ICT supporting creativity in Chapter 6.

Chapter 2

Assessing mathematics with ICT

In this section you will:

■ consider how ICT resources can support assessment for learning in mathematics;

■ look at some ICT tools that help you develop effective questioning techniques;

■ preview some innovative ICT tools that allow you to 'listen' to pupils' learning.

Opportunities for pupils to express their understanding should be designed into any piece of teaching, for this will initiate the interaction whereby formative assessment aids learning.

Paul Black and Dylan Wiliam *Inside the Black Box* 1996, page 11

To begin with, I should be clear about how I am choosing to interpret the meaning of the word 'assessment' within this book. Some ICT resources, such as Integrated Learning Systems (ILS), claim to raise pupils' attainment through a pre-determined programme of individualized learning. But I do not believe that the items of such programmes are effective as *learning* activities, or efficient uses of ICT within mathematics. There may be a few pupils who are close to 'understanding' a particular calculation, procedure or technique, for whom the experience is a support in enabling them to feel some success. However, the value of using ILS systems as an *integral part* of the learning process is somewhat dubious.

You need only review the evaluation criteria introduced at the end of Chapter 1 to see how few of these ideals are met. Within any individualized assessment programme, there should be opportunities for pupils to:

➡ communicate mathematics appropriately, for example make informal jottings, make sketches or answer in words;

➡ make relationships across and within different mathematical concepts;

➡ make estimates first, and then refine answers;

➡ simplify a task or problem that is difficult;

➡ be imaginative, creative or flexible.

The arrival of the Tablet PC does mean that we are moving towards being able to input handwritten responses, which could include sketches, diagrams and comments. Even so, the ideal 'listener' to these responses is a teacher, not a computer!

I am more interested in exploring how ICT tools can be used effectively to support teachers in assessing pupils' understanding as a route to enhancing their learning.

In their booklet, *Assessment for Learning: Beyond the black box*, the Assessment Reform Group have described what assessment for learning means to teachers in practice:

'... teachers must be involved in gathering information about pupils' learning and encouraging pupils to review their work critically and constructively. The methods for gaining such information are well rehearsed and are, essentially:

➡ observing pupils – this includes listening to how they describe their work and their reasoning;

➡ questioning, using open questions, phrased to invite pupils to explore their ideas and reasoning;

➡ setting tasks in a way which requires pupils to use certain skills or apply ideas;

➡ asking pupils to communicate their thinking through drawings, artefacts, actions, role play, concept mapping, as well as writing;

➡ discussing words and how they are being used.'

Designing lessons and teaching activities that have the assessment of pupils' mathematical understanding at their heart is no mean feat!

Irrespective of what is known about pupils' prior knowledge in the form of data or 'traffic-lighted' assessment records, it is not until you begin the lesson or activity that you start to unpick what pupils already know, understand and can do. The aim should be to choose activities that enable pupils to review and connect the ideas with what they already know, and develop their thinking by challenging them to become actively engaged in some new mathematical ideas and strategies.

How many times have you carefully planned a lesson, using appropriate yearly objectives from the relevant Framework for teaching mathematics, only to find that the pupils' reactions suggest they have never met the topic before? Or, conversely, they effortlessly glide through the task with no new learning taking place.

More and more teachers are using the early part of the lesson to assess what pupils already know about a topic, in advance of beginning their planning. That is, they may be preparing the ground for the introduction of a new topic in the coming weeks.

A KS3 teacher, in order to assess what a lower ability Year 7 class already understood about fractions, decimals and percentages, used the following example. She chose an image from Richard Phillips' Problem Pictures CD-ROM, which she displayed through a data projector onto a Smartboard®. A Smartboard® is the type of interactive whiteboard on which you drag objects and annotate using your finger.

She chose the following image:

Initially, she asked the pupils, 'What do you see?'

As the school operates a 'no hands up' policy, the immediate response from the pupils was to turn to their response partner, and a wave of discussion swept around the room. The teacher listened intently to the discussions, picking out key mathematical words as pupils used them, and noting them on a separate traditional whiteboard mounted to the side of the electronic one.

After about a minute she brought the class back together and invited individuals to respond to the question. Responses included 'a chocolate bar', 'just over five pieces of chocolate unwrapped' and 'it is a rectangle shape'.

The teacher then probed the pupils' thinking further by asking 'If we were to fully unwrap the chocolate bar, how many pieces of chocolate would you estimate that there are altogether?'

Again, she did not expect 'hands up', or take an immediate answer from one pupil. Instead she gave the pupils time to think, before asking them all to make any diagrams or jottings on their individual whiteboards and record their individual response. After about another minute, she invited a pupil to hold up his responses and explain how he had arrived at it. She then asked if any pupils had a different answer, and gave them the opportunity to explain their thinking. At no time did the teacher suggest that she was expecting an automatic 'right answer'. Instead she placed the emphasis on pupils justifying their individual responses, praising their mathematical reasoning and correct use of vocabulary.

The teacher then asked pupils to think about how they might draw some lines over the top of the picture of the chocolate bar, to help decide how many pieces of chocolate composed the whole bar.

One pupil used his finger to annotate the diagram, producing something similar to the image below.

At this point the teacher asked pupils how they might estimate, as a fraction or percentage, how much of the bar was unwrapped. Pupils were happy with 'less than a half' and 'less than a quarter'. The teacher asked how they might obtain a better estimate based on their knowledge that there were 32 pieces of chocolate in the bar.

The teacher now turned to a different ICT tool, Research Machines (RM) 'Easiteach Maths®' (v.2), which is described in more detail in Chapter 4.

In the table following, the left-hand column shows the images that the teacher had prepared before the lesson, the middle column gives the questions that she asked to probe the pupils' understanding, and the right-hand column gives the final images.

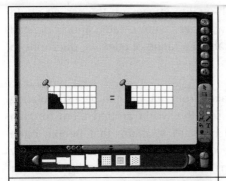

Do you agree that these two images show the same fraction shaded?

Can you describe how it has been done?

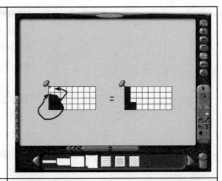

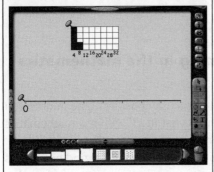

Look carefully at the number line. Tell me what you see.

How do you think we could use this number line to help us estimate the fraction of the chocolate bar that was unwrapped?

The teacher also held up a 4 by 8 by 1 cuboid of brown Multilink Cubes that represented the complete chocolate bar. She gave this to a pupil to support her in explaining the idea that the bar could be broken into 8 sections, each containing 4 pieces.

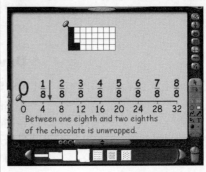

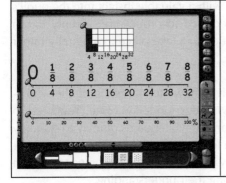

Look carefully at the new number line on the screen.

What does it show?

How could we use it to estimate, as a percentage of the whole chocolate bar, the part that was unwrapped?

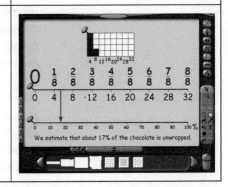

The final images capture the pupils' responses to the questioning; they do not represent a set of 'aims' for the teaching. However it is reasonable to assume that, by reviewing previous knowledge, vocabulary and representations, many pupils would have been making connections in their own minds. This is an example of an aspect of 'assessment for learning' in practice. By focusing on assessing what pupils already knew and understood, the teacher has been able to offer them an active learning experience. She also has a good idea of how the learning could be developed. In this particular example, the teacher used ICT-generated visual images effectively to support models and images that some of the pupils already held in their heads. In doing so, other pupils were able to 'see' how such images help us talk about mathematics.

This assessment activity took no longer than five minutes. But while doing it the teacher needed to have a clear idea of what it was that she wanted to assess, and how she could use features of this particular ICT tool. It is easy to see how the availability of a range of clear, high-quality images on the screen could support pupils. The ease with which new images were created and/or deleted in direct response to pupils' thoughts enabled a range of effective questions to be asked within a short space of time.

This activity also shows how an image taken from real life can be a stimulating starting point. It encourages pupils to 'see' the world around them through mathematical eyes. Although it would be hard to claim that people often engage in fraction calculations while devouring bars of chocolate, pupils may begin to ponder mathematical questions for themselves in the light of their experiences in the classroom.

Developing effective questioning in the mathematics classroom

Asking effective questions can be perceived as a risky business. The minute you begin to ask pupils 'How…?' 'Why…?' and 'What if…?' within a mathematics lesson you need to be ready for the wide range of responses that you will hear. And it is very important to listen, really listen, to the responses, finding aspects of pupils' responses to praise. Think back to your own experience of learning mathematics at school. Many of you may remember occasions when you were, or a fellow pupil was, completely deflated or humiliated by an insensitive teacher who mocked a wrong answer.

Every wrong answer masks a miscalculation, a misinterpretation or a deep-rooted misconception, and provides you with a valuable insight into the pupil's thinking. Often further questioning is required to probe more deeply into a pupil's thoughts, although such intervention can take place individually rather than in front of the whole class.

Many teachers are finding that using individual whiteboards during the whole-class teaching phase is invaluable in giving every pupil a voice. You need to ensure that you are actively listening. Scan the whiteboards quickly to spot any surprising responses. Make sure that you follow these up, or note them in your lesson evaluation. You can then plan how you are going to intervene to support the particular pupil or pupils in developing their understanding.

When you begin to listen to how others describe mathematical ideas and strategies, you may begin to question your own understanding. In Chapter 3 we will focus on visualizing ideas in mathematics, and look at examples in which, in various contexts, pupils and teachers are describing what they see. The nature of assessing learning by listening requires you to have an open mind, and may challenge your mathematical upbringing. Were you brought up to see mathematics as a set of rules and procedures to be rote-learned and replicated in tests? For those of us who experienced such a mathematical diet it is enlightening to revisit mathematics within a more open culture. Relax your guard and actively encourage your pupils to develop more positive attitudes to the subject. When questioned, pupils often reveal that they see the teacher as the gatekeeper to the knowledge in mathematics classrooms. For them the means to success is through ticks and crosses on the page, and you, the teacher, issue these!

Evidence from the assessment for learning research clearly indicated that, when the assessment feedback was comment-based, this perception by pupils changed for the better; the comments opened a dialogue between pupil and teacher that focused on the mathematics.

So an effective question is focused toward the process of learning mathematics, not the answer. That doesn't mean that you never actually ask pupils for answers, but that you are likely to follow up a 'closed' question such as 'What is 23 per cent of £150?' with a probing question such as 'How did you work that out?'. An alternative approach would be to begin with 'How would you work out the answer to 23 per cent of £150?' followed by 'And what is the answer?'.

What is the difference between these two approaches? In the first approach, the emphasis was at first placed on the answer, reinforcing some pupils' views that mathematics is 'all about right answers'. The follow-up question asks one pupil to expand on their own strategy, which may not fully engage the ears of the rest of the class.

The second approach asks all pupils in the class to consider how they would work out an answer. Each pupil would have an approach, even if it was incorrect, and you will have engaged them all in thinking mathematically about the problem. The follow-up question asks them to put their approach into practice; you might use individual whiteboards so that all pupils can offer an answer. By asking about the strategy before the answer, you have prepared pupils to think about how they would solve the problem before they are asked to find the solution.

Even if our only aim was to prevent pupils hurriedly writing down wrong answers in standard assessment tests the second approach would be better, encouraging them to think about their strategy first before launching 'blindly' into calculations or procedures and might help pupils to become more successful when answering test questions!

A rich question to ask pupils about a diagram, graph, or other visual image, is 'What do you see?'

No matter what the response, you cannot argue with the pupil's answer!

Take the following diagram – what do *you* see?

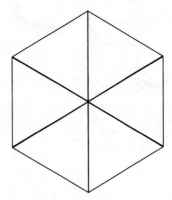

Some possibilities are:

A tessellation of six equilateral triangles.

A playing piece from Trivial Pursuits game.

The skeleton of a cube.

Mundhar Adhami of King's College, has written, in *Support for Learning*, about 'responsive questioning' and says:

'It seems possible to regard responsive questioning as the main and essential quality of good teachers. But there seems no easy route to develop such skills.

Three main requirements for responsive questioning can be identified:

1. Flexible knowledge of the activity and topics involved, including common misconceptions and the different connections that can be made from lower order to higher order concepts.

2. Genuine acceptance and engagement with pupils' ideas in their own language. Based on (1) above these ideas can then be sifted and focused while pupils retain ownership and move in their thinking.

3. A sense of the "optimum" or the "best possible" decisions in a given classroom situation. Every class and every lesson is different, and teaching involves continuous adjustments on response to the unfolding scene. The teacher must engage with various contributions, at the same time keeping an eye on the overall agenda, and keep the momentum, and be aware of the time. A good lesson is where most pupils have met personally meaningful challenges and moved in their thinking.'

Anne Watson and John Mason have developed an excellent resource book for teachers, *Questions and Prompts for Mathematical Thinking*, published by the Association of Teachers of Mathematics. They suggest a range of starting points when structuring questions for specific purposes:

Exemplifying, Specializing	Completing, Deleting, Correcting	Comparing, Sorting, Organizing
Give me one or more examples of…	What must be added removed altered in order to allow…? ensure…? contradict…?	What is the same and different about…? Sort to organize the following according to…
Describe, Demonstrate, Tell, Show, Draw, Find, Locate, an example of…		
Is … an example of?		Is it or is it not…?
What makes … an example?	What can be added removed altered without affecting…?	
Find a counter-example of…		
Are there any special examples of…?	Tell me what is wrong with…? What needs to be changed so that…?	

Changing, Varying, Reversing, Altering	Generalizing, Conjecturing	Explaining, Justifying, Verifying, Convincing, Refuting
Pre-cursor - Alter an aspect of something to see effect. What if...? If this is the answer to a similar question, what was the question? Do... in two (or more) ways. What is quickest, easiest, ...? Change ... in response to imposed constraints.	Of what is this a special case? What happens in general? Is it always, sometimes, never...? Describe all possible ... as succinctly as you can. What can change and what has to stay the same so that ... is still true?	Explain why... Give a reason... (using or not using...) How can we be sure that? Tell me what is wrong with... Is it ever false that...? (always true that...?) How is ... used in ...? Explain role or use of... Convince me that...

If you consider the richness of these question structures and then look inside a typical school mathematics textbook it is easy to see how opportunities to engage pupils in real mathematical thinking can be missed.

ICT can greatly enhance the presentation to pupils of mathematical models and images that offer rich starting points for mathematical activity in which some of these questions may be asked. Throughout this book I will offer a range of ICT resources that particularly accommodate themselves to such a questioning environment.

Developing a questioning culture in the classroom

If you are incorporating more ICT into the classroom, which is requiring pupils to think more deeply about mathematics, and you are using questioning to develop their understanding, you may need to develop a different classroom culture.

Let's consider what has been written about the social culture within successful classrooms.

Alistair Smith in *Accelerated Learning in Practice* says that the following factors are essential for developing pupils' positive self-esteem and accelerating their learning experiences:

- ➡ an overriding sense that the student is part of a group and their contribution, whatever its nature, is valued – they feel belonging;
- ➡ students are encouraged to set and work towards their own achievable goals and reflect on their progress as they do so – they are working towards their aspirations;
- ➡ the classroom and the learning environment are safe havens where there is consistency in expectations and standards – they experience safety;

➡ a realistic level of self-knowledge is supported by the belief that individuality is not threatened by undue pressure to conform – their identity and individuality are recognized;

➡ mistakes are valuable learning tools in an environment where one can take risks and achievement is valued – the learner achieves success.'

More specifically to mathematics classrooms, James Hiebert and others, in their book *Making Sense: Teaching and learning mathematics with understanding*, conclude that, in a good environment for learning mathematics, the social culture should be one in which:

' ➡ ideas and methods are valued;

➡ students choose and share their methods;

➡ mistakes are learning sites for everyone;

➡ correctness resides in mathematical argument.'

It is the role of the teacher in the classroom to establish such a social culture and, ideally, the department and the school as a whole would share the culture. But this culture cannot be developed in isolation from other aspects of mathematics teaching and learning. As the classroom teacher, you decide on the tasks the pupils are going to do. If the chosen task is an exercise of repetitive questions, how will you ensure that the desired culture is achieved? Would you need to rethink the nature of the classroom tasks in order to achieve the desired classroom culture? And what about the resources that you might choose to support pupils' experiences? After all ICT is just another tool that is available to support teaching and learning.

If establishing the social culture cannot be considered independently from other aspects of the mathematics classroom, what are the aspects? Hiebert and his colleagues usefully define them as the 'five dimensions' of the classroom, which, in no particular order, are:

➡ the nature of classroom tasks;

➡ the role of the teacher;

➡ the social culture of the classroom;

➡ the mathematical tools as learning supports;

➡ equality and accessibility.

When adults are asked about their own experiences of learning mathematics at school, irrespective of the level of mathematics they attained, there are very few who describe it as a positive experience. A few can cite individual teachers who influenced them in a positive way, and when asked to expand on this, it is often the culture of their classrooms which distinguished them from other teachers.

Alistair Smith, in *Accelerated Learning in Practice,* summarizes recent brain research into learning. On the topic of challenge and stress, he says:

'The brain responds best in conditions of high challenge with low stress, where there is learner choice and regular educative feedback. Multi-path, individualized and thematic learning with the mental work of engaged problem solving enriches the brain. So does novel challenge and real-life experience.'

It could be argued that mathematics, which is often perceived by pupils to be the hardest subject that they study in school, has the furthest to go in terms of improving aspects of how it is taught and learned. If it is, by its nature, a challenging subject, then the key role for teachers is to make the process of learning mathematics as stress-free as possible.

Many mathematics schemes try to do this by 'taking the nasties out'; that is, by providing repetitive sequences of questions that do not give pupils a reason to challenge their understanding, or tackle real problems. Hence pupils do not actually move out of their comfort zone and will compliantly complete text-book exercises. Then they say that mathematics is boring!

How can ICT support effective questioning?

I see providing some answers to this question as an essential aim of this book. After all, if we are talking about ICT enhancing learning and we accept the importance of asking effective questions, then the obvious next step is to put the two together.

So, let's begin with a simple example.

The screen below shows a simple program called 'Broken Calculator'. It is freely available as one of the applets on the website Wisweb, which is a project co-ordinated by the Freudenthal Institute in Holland. (http://www.fi.uu.nl/wisweb/welcome_en.html)

WEBSITE

The ⬚ key represents division. Might this suggest that Dutch children are better at making connections between division and ratio?

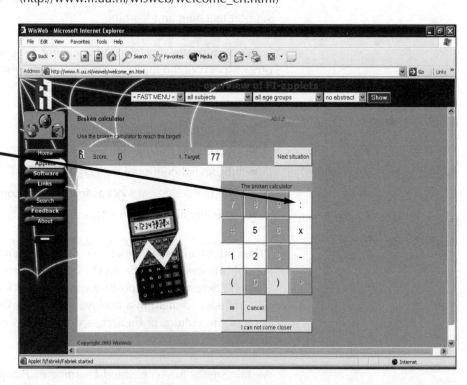

The task is to try to make the target number, or get as close as you can to it, using only the highlighted keys.

It would be easy to picture a class of pupils in an IT suite spending some of the lesson time working through this activity and other similar activities on this website. But, would using a data projector to display just one broken calculator to the whole class provide a richer learning experience? You could then ask the whole class questions such as:

Using only the keys that are allowed, do you think that it may be possible to make the target number?

What makes you think that?

Do you think that there may be more than one way?

What if you are not allowed to use the brackets?

What if you have to try to use every key that is allowed?

How many different ways can you reach the target number?

Can you design your own broken calculator and choose some target numbers that you think would be easy to make... or impossible to make?

This is a simple ICT resource, and there are a host of questions that could be asked in order to develop pupils' mathematical thinking. Why not try it with your pupils? As a staff professional development exercise, you could take another simple ICT resource, and devise a list of questions to ask pupils.

The future of assessment of pupils' ICT-based work

Pupils may be engaged in ICT-based activities that do not require them to record their work on paper, or the software may not allow them to make notes on-screen. There are some Shareware packages that can be downloaded from the internet which allow you to video record the computer screen. If a microphone is attached to the computer, the pupils' discussions can be recorded too. One such package is CamStudio, a free download from

WEBSITE

http://www.brothersoft.com/CamStudio_Download_3944.html.

Finally, in looking to the future, should we also think about the changing nature of assessment?

In his book *Multiple Intelligences: The theory in practice*, Howard Gardner defines assessment as:

'the obtaining of information about the skills and potentials of individuals, with the dual goals of providing useful feedback to the individuals and useful data to the surrounding community. What distinguishes assessment from testing is the former's favouring of techniques that elicit information in the course of ordinary performance and its general uneasiness with the use of formal instruments administered in a neutral, decontextualized setting.'

If we are embracing Howard Gardner's 'multiple intelligences' theory, as indeed many schools are, the role that ICT can have within this broader view of assessment is evolving. The development of the use of ICT as a tool for testing is missing the real opportunity for ICT to develop as a tool for teaching and learning. I would urge those who are currently developing on-line assessment packages to consider carefully their rationale for doing so. Yes, assessing incorrect/correct answers to straightforward mathematics problems may be cheap and simple to program, and welcomed by teachers because it reduces marking and record keeping. But is this an expensive way to use resources that can be better employed for the real benefit of pupils and teachers?

Chapter 3

Visualizing mathematics with ICT

In this section you will consider:

■ what is meant by visualization in mathematics and how it supports learning;

■ the role of mathematical tools in relation to visualization;

■ some technological tools that are available to support visualization.

▌ What do we mean by visualization?

Every human being has a very individual way of visualizing mathematical concepts and processes. Developing mental models and images can be a great help to us in understanding mathematics.

For example, it is possible to describe to somebody in words the geometrical object that we call a square. However, the description makes more sense if there is a diagram with it. Having established a visual image of a square, if somebody talks about the square or moves it about, we can mentally hold the properties of the square in our heads. Our square keeps its four equal sides, four right angles, four lines of symmetry and rotational symmetry order four, even if we transform it mentally in some way.

Try the following visualization activity (see overleaf):

Imagine an isosceles triangle...

Focus on one of the sides of the triangle and let it become a line of reflection. Reflect the rest of the shape in this line to form a new shape.

How many sides does the new shape have?

Can you name your shape?

Without drawing your shape, describe your new shape to a partner?

Have you both described the same shape?

It is easy to see how examples from shape and space provide opportunities for us to visualize mathematical ideas. However, how often do we consider visualizing other areas of mathematics?

Think about the following calculation: 15 x 27

You may well be able to calculate the answer, but in how many different ways can you visualize this calculation?

Do you visualize it as an array of dots, arranged as 15 rows of 27 dots?

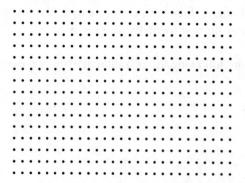

Or as a 15 by 27 rectangle of squares?

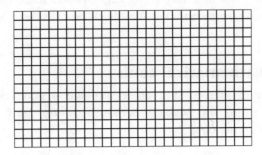

Does the way that you visualize the calculation help you to work out the answer?

Did you mentally partition the numbers into tens and ones to perform $(10 + 5)(20 + 7)$?

Or did you see it as $15 \times 30 - (15 \times 3)$?

If you ask colleagues or pupils these questions they are likely to reveal a range of different responses, both in terms of the mental pictures and the mathematical vocabulary they choose to use.

Alternatively, how do you visualize the calculation $3 \div 4$?

A group of primary teachers on a professional development course in Croydon LEA were asked this question. The responses were varied and interesting:

➤ Alan and Sarah said that, as they both had prior knowledge that $3 \div 4$ is an alternative way of writing $\frac{3}{4}$, the question was trivial and they didn't understand what was meant by 'visualizing' the calculation.

➤ Annette talked about visualizing 3 pizzas, each divided into 4 parts. She imagined each person having a quarter of each pizza, giving each person 3 quarters of a pizza.

➤ David imagined a number line, from zero to 3, and halved the line twice to arrive at the position 0.75. He gave the answer as a decimal.

Abraham Arcavi (2003) has written on the topic of mathematical visualization and he proposes the following definition:

'Visualization is the ability, the process and the product of creation, interpretation, use of and reflection upon pictures, images, diagrams, in our minds, on paper or with technological tools, with the purpose of depicting and communicating information, thinking about and developing previously unknown ideas and advancing understandings.'

Arcavi's definition lets us concentrate on the purpose of visualization within our context of developing mathematical understanding. What we are interested in is how technological tools, as he calls them, can support this process.

Mathematical tools

James Hiebert and others provide a good definition of tools to support learning in *Making Sense: Teaching and learning mathematics with understanding*.

'Tools are resources or *learning supports* and include skills that have been acquired, physical materials, written symbols and verbal language.

The role of tools is both to record and communicate mathematics and to support us to think.'

I like this broad definition of tools as it encompasses *all* the resources that are at our disposal when we approach a new mathematical problem. The role of the teacher is to plan how to maximize the use of these resources in their teaching.

How often do we plan lessons that really take account of what pupils already know and can do?

Do we really understand how the physical resources support the learning, and how a subtle change of resource can change the learning experience and the outcomes?

Hence most good lesson plans will include a consideration of what the pupils already know, which practical resources are required and which words and symbols will be used or introduced.

If we focus on the physical materials or practical resources that have been developed to support mathematical learning, the interesting question is *how* do they work?

Many specific mathematical tools have been developed to support us to visualize mathematics:

Place value chart invented by Caleb Gattegno

Base-10 blocks invented by Zoltan Dienes

Number rods invented by Georges Cuisenaire

Resources such as squared paper, isometric paper, interlocking cubes and polygons are all commonplace in mathematics classrooms. Each of these tools was designed to support us to foster understanding by conceptualizing a mathematical idea or ideas.

James Hiebert and others make the following points about the roles that tools play when we are learning mathematics:

'Mathematical meaning must be constructed for, and with, the tools by the user. Different types of meanings can be developed from the same tools.'

Try the following experiment with some pupils, or consider what your pupils might do:

Provide pupils with squared paper and a pencil, but don't set a task! Tell them that they can do whatever they like!

What do they do?

Do they have an expectation of what you might expect them to do – if they associate you with mathematics do they draw shapes, or a graph?

Or do they draw patterns or ask for coloured pencils?

Or do they fold the paper and not use the pencil?

Or do they just write on the paper?

A piece of squared paper can be a powerful mathematical tool within the context of a task. On its own it is whatever the user wants it to be!

Annette Johnson, a primary teacher, working on developing the use of 'learning diaries' with Year 6 pupils in a Croydon school, inadvertently explored this. The school had run out of the lined exercise books that the pupils had used previously and, instead, provided 0.5 centimetre-squared books from an abundant stock. The way in which the pupils used their learning diaries changed significantly. The pupils began to make comments related to work they had been doing during numeracy lessons. They invented mathematical games, they drew reflection patterns, practised their developing numerical skills and began to pose their own mathematical questions.

If you want the tool to support mathematical understanding, it is part of your teaching role to decide how the tool is going to be used *and* how the pupils are going to interact with it. It is not enough to simply give out the tool.

Visualization and technology

Paul Goldenberg writes, in *Issues in Mathematics Education*:

'With technology, what changes is the pool of problems to choose among and the ways they can be presented... Some lessons require pupils to experiment with certain mathematical objects and see how they respond. Some require visual representations – graphs, diagrams, geometric figures, moving images – that respond to students' questions, answers or commands.'

The Association of Teachers of Mathematics have developed a range of resources on the CD-ROM called Developing Number, and one of these is called 'Number'. It provides an interactive Gattegno place value chart that talks! One of the key ways in which 'Number' supports pupils' developing understanding of place value is through the combination of the Gattegno chart, saying and hearing the number names, and the dynamic images as the numbers are built.

Dave Hewitt, of the University of Birmingham, has written about some of the key design features of this suite of software in *Teaching Secondary Mathematics with ICT*.

He writes,

'Place value is often seen as a significant topic and it can become more of an issue when working with decimals. A number such as 384.87 is often read aloud as three hundred and eighty four point eight seven. Here, the digits after the decimal point are not given value names, it is just said as ...point eight seven. A second issue is that, particularly within the context of money, students can become used to saying ...point eighty seven, thus having the same value name for both the eights. The software "Numbers" has the possibility of having the value names extended into the decimals with additional rows given names tenths, hundredths and thousandths.'

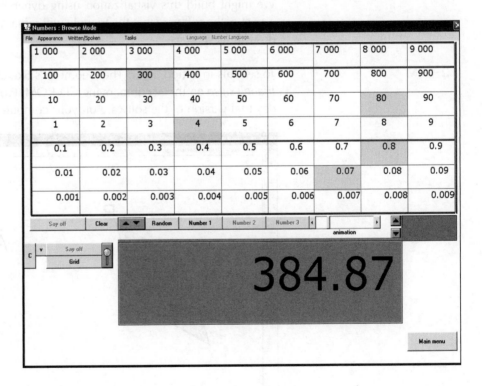

WEBSITE

The 'Developing Number' software is available from the Association of Teachers of Mathematics website (http://www.atm.org.uk/resources/developingnumber.html).

Why would you consider using the software in addition to using a Gattegno tens chart poster in the classroom?

A clear advantage is the way in which using the software links the visual, kinesthetic and auditory learning of place value.

The visual image of the numbers sliding into place from their position in the chart, as well as the facility to hear the values of individual digits, is not immediately evident from the tens poster. The software supports learners to construct their own mathematical meanings for the tens poster by providing additional sensory stimuli.

Think back to the visualization activity with the triangles at the beginning of this chapter. How could ICT have supported this activity?

> ...Imagine your isosceles triangle that you have reflected to produce a quadrilateral...
>
> Now imagine dragging one of the vertices (corners of your shape). Remember, your triangle must remain isosceles... What happens now?

I don't know about you, but my mind finds it hard to move one of the vertices *and* keep my original triangle an isosceles one.

We might build this visualization using dynamic geometry software. It is an aspect of the software that the triangle will remain isosceles, no matter what we do with its vertices.

CD–ROM

To see this in action, open The Geometer's Sketchpad file *Reflecting isosceles triangles.gsp* on the accompanying CD-ROM. (You may need to install the 30-day trial version of the software on your computer first.)

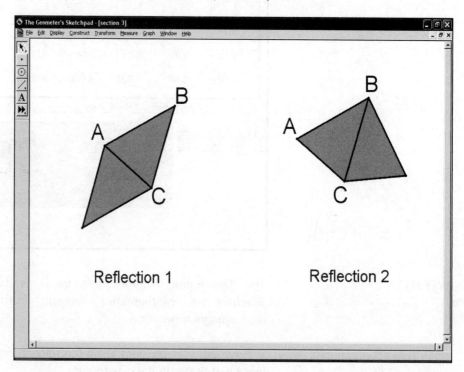

The original triangle is labelled ABC.

Two different reflections are shown, the first uses the line AC as the line of reflection and the second uses the line BC.

Which side did you choose as your line of reflection?

Use your mouse to drag any of the vertices A, B or C on either reflection.

What happens?

Can you say what is staying the same and what is changing?

Which quadrilaterals is it possible to make?

Which quadrilaterals is it not possible to make?

Immediately, we have created a real-life environment in which we can play. And the ICT is allowing us to play with the properties of the shapes, not just with the set of named shapes.

ICT resources to support visualization in division

In my experiences when working with teachers in both primary and secondary schools, division is always among the 'top ten' topics with which teachers perceive pupils experiencing difficulty. So how can ICT help?

The National Numeracy and Key Stage 3 Strategies have developed a number of Interactive Teaching Programmes, which are animations programmed in Flash with a range of interactive features. These are available as free downloads from the standards site, **WEBSITE** http://www.standards.dfes.gov.uk/numeracy/publications, and can be saved to your school's network for easy access.

To support pupils to develop their mental imagery, some of the ITPs use mathematical imagery, such as a number line, alongside other representations.

An example of this is the ITP 'Groupings'.

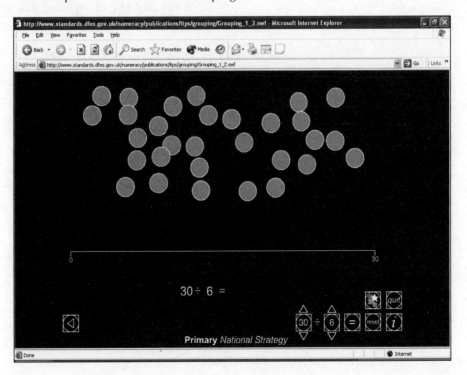

The range of on-screen tools, which have been designed with interactive whiteboards in mind, enable the teacher or pupils to access immediate visual support for a division calculation, in this case 30 ÷ 6.

For example, clicking on sets of 6 counters results in the following screen.

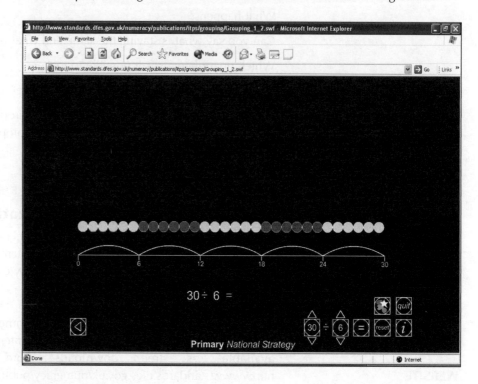

On first appearance, this type of ICT resource focuses on a specific area of mathematics, in this case the notion of division as the process of grouping. However, think back to Chapter 2, in which we looked at effective questioning. The use of this resource could be enhanced by asking questions such as:

'What could be changed so that the answer is still 5?
... and what would the new screen look like?'

or

'If the answer to a similar question is 4, what could the question be?
... and what would the new screen look like?'

In both questions the pupils are being asked to conjecture an answer, and pupils will have individual reasons for their choices. So, do you allow all pupils to 'speak' by giving them individual whiteboards and pens to show their thinking or do you have a half-class set of laptops that the pupils can use to test their conjectures?

Remember, the most important consideration is not what the ICT tool can do, but *how* you, as the teacher, choose to use it!

Staying with the topic of division, a software package 'Division Exploded' really does attempt to explode the concept of division and, at the same time, offer teachers the opportunity to support pupils' progression from mental to written calculations.

You only need to look at the Qualifications and Curriculum Authority's feedback on pupils' performances in the mathematics SATs at the end of Key Stages 2 and 3 to see that many pupils have difficulties recording their work when faced with division problems.

'Division Exploded' is another ICT resource, designed by John McCormack, which allows you to explore different representations of the same calculation and switch screens to move between them.

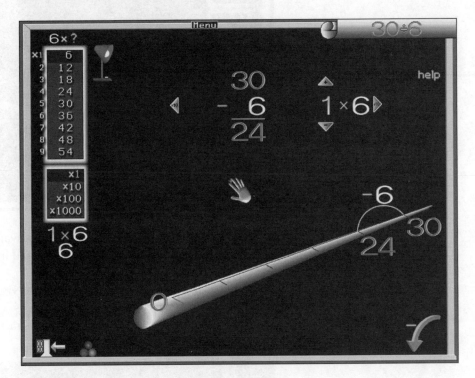

The above screenshot shows the calculation 30 ÷ 6 in 'number line' mode with the repeated subtraction calculation turned on (by clicking the light).

The amount of sixes to be subtracted can be altered by using the

 or

icons, dragging the position on the number line itself or selecting a 'chunk' to subtract from the table of multiples of 6.

The division calculation can also be approached as a repeated addition from zero by clicking the [icon] icon.

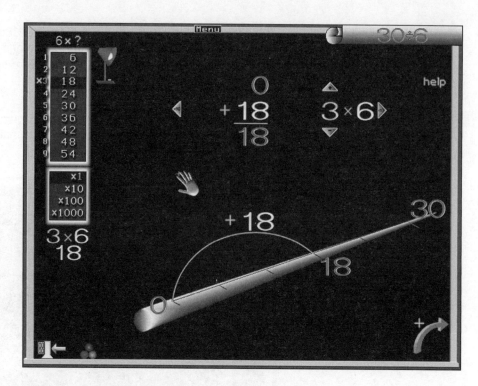

CD–ROM

Division Exploded is definitely a piece of software which needs to be played with, so load the accompanying CD-ROM and install the 30-day trial version.

The number line images encourage valuable visualization in pupils across all Key Stages. However, the focus in Key Stages 3 and 4 is very much towards developing an efficient written calculation strategy for division. So, choosing some larger numbers, 353 ÷ 17...

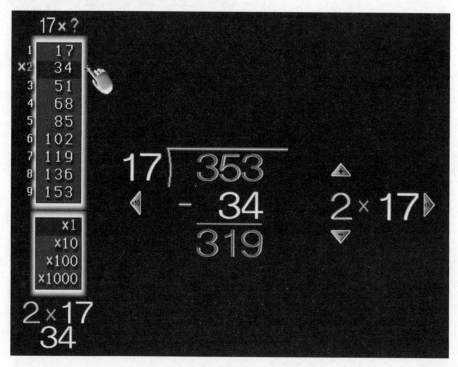

The software allows you to subtract successive multiples of 17. However, as the dividend is greater, it also allows you to multiply the chunks by 10 using the [icon] icon.

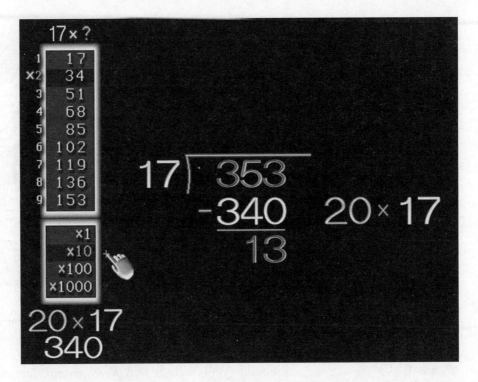

This is leading pupils towards a more efficient strategy and supporting them to gain a real understanding of how the standard written method works. It links the standard written method to the concept of chunking, which was introduced through practical activities and number line work in Key Stage 1, and developed further in Key Stage 2.

The dynamic and exploratory nature of this particular ICT tool is begging pupils to play with numbers. If this software is the chosen tool, consider how the additional imagery would support pupils' understanding of division over and above being able to perform the calculations correctly in examination situations!

▌ Using visual images in the classroom

So far we have focused on the role of visualization within the learning of mathematics. We also need to consider how we can incorporate mathematical images.

The wealth of digital images available from the internet, and some creative use of existing software, open up a whole new way in which we can interact mathematically with images. A simple example of this was given in Chapter 2 using the chocolate bar image.

In my mathematics classrooms I always had bright stimulating posters, many of which were obtained from Tarquin Publications. One of my favourites is a beautiful poster 'Animal Tilings' by Andrew Crompton, which features Escher style animal tilings such as the Elephants and Hedgehogs.

Some years later, with the world wide web at my fingertips, an 'I wonder' search led me to discover that Andrew, who lectures in Architecture at the University of Manchester, has put images from the poster, and many more beside, on his website http://www.cromp.com/tess/home.html. How could we incorporate these images into mathematics lessons and bring in ICT?

WEBSITE

The first stage of 'catching' images from a website is to hold your mouse over the image and 'right click'. This reveals a menu from which you can choose whether to print straight away to a connected printer or copy, save or email the file to use elsewhere. The size of the image will have been fixed on the web page so you may need to adjust this before you print.

In this way it is easy to obtain good quality copies of the images that could be laminated for use in the classroom. But you need to be aware of copyright restrictions if you are making multiple copies.

In Key Stage 2, pupils could explore the symmetry of the tessellations using tracing paper and card cut-outs. They may begin to require the vocabulary of symmetry and transformation in order to describe what they find out.

Up to this point the ICT has been used to obtain and create resources to support traditional teaching approaches. But this resource can be taken a big step further. One of the most exciting ways to use digital images is to import them into dynamic geometry software and explore transformations that produce a tessellation.

Begin by pasting or importing onto the page, a digital photograph such as this one (left).

Just look at the image. Ask pupils what they see.

A process that is fundamental to being a mathematician is *conjecturing* and *testing* mathematical ideas. This can be likened to mathematical play.

What do we conjecture about the flying fish tessellation? An early observation is that it appears to have something triangular, or hexagonal, about it.

As we have chosen to use dynamic geometry software, we can mark points and join them with line segments to highlight the symmetry within the animal tiling. If you were using an interactive whiteboard, you could annotate over the top of the software.

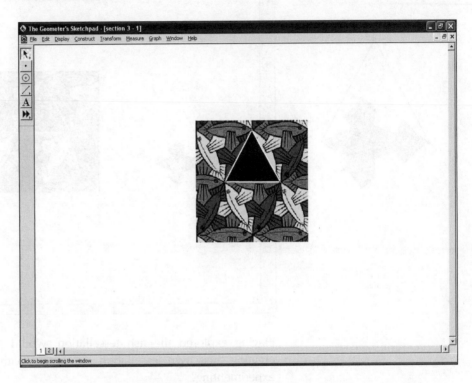

In the screenshot above, three points have been drawn and joined to form what appears to be an equilateral triangle.

Using the software we can choose a point to become a centre of rotation and rotate the whole triangle to create a second one.

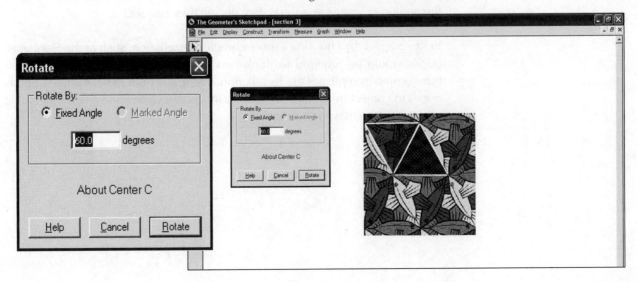

We could then look closely at the fish shapes. What do you conjecture?
... that each fish is rotated 3 times by its head and 3 times by its tail...
... that, ignoring decoration, each fish has a line of symmetry along the length of its body...
... that 3 fish fins meet mysteriously at the centre of the triangle...
... or something else?

To help test their conjectures, pupils can put points, line segments and arcs onto the diagram, and use the software to transform them.

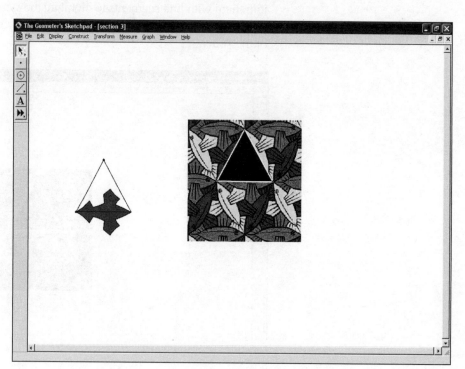

Having explored the fish tessellation, would pupils begin to think about creating their own tessellation by starting with an equilateral triangle and experimenting?

Given three folded pieces of card, could pupils now begin to think about how to design their own animal tiling? They may discover that the key to designing a tessellation of the flying fish type is to look at one very small part of the tessellation, that repeats itself all over the tessellation, and, in doing so, gives the outline of whatever animal they decide they can see!

In Key Stage 3, by choosing a more complex tessellation, such as the Elephants, pupils would be required to think how that tile might be created and then transformed to replicate the tessellation. Feedback from the software supports pupils to conjecture and test their ideas in a non-judgemental environment that allows them to make fast progress.

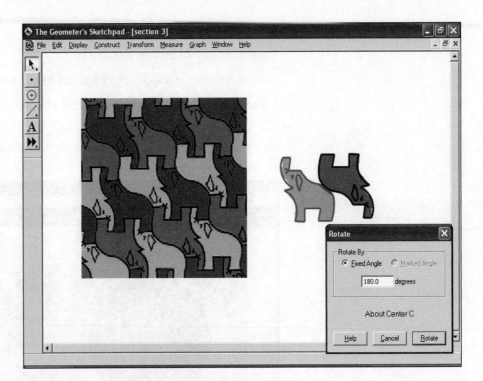

It would be easy to ask who is actually teaching the pupils. In the previous examples the software would appear to be taking a lead role. But do remember that it is the teacher who has designed the task, decided how and when pupils access the ICT, focused the pupils towards a clear learning outcome and is far from redundant!

Can ICT support visualizing in 3-D?

The Royal Society report, *The Teaching and Learning of Geometry 11–19*, published in 2001, made the following comment.

'The working group would like to see further development of the curriculum with respect to work in 3-D and the use of ICT.'

In some respects, when we live in a 3-D world it seems odd that we might consider using the computer screen, a 2-D world, as an environment to explore a 3-D world. How can we possibly get a feel for the nature of 3-D shapes and their properties? But the implications for science and industry are far-reaching. Within any CAD-CAM software environment, the computer operator is making design decisions based on what they see on the screen. Misinterpretations and miscalculations become expensive manufacturing errors! Pupils are beginning to access this type of technology within the design and technology curriculum, an area within which mathematics has a key role. It is worth visiting a school with a design and technology department that has developed CAD-CAM work to see how the mathematics curriculum overlaps. Cross-curricular work, based on the pupils' own experiences, can grow from an awareness of the common areas of mathematics that exist.

WEBSITE

Recently, I came across a website, developed by the Utah State University in the United States (http://matti.usu.edu/nlvm/nav/vlibrary.html), with the snappy title 'The National Library of Virtual Manipulatives'! Being curious, I selected Grade 9–12 Geometry, chose the file 'Platonic Solids' and started to play.

I chose a dodecahedron and started to roll the shape, using the mouse.

I could enlarge or reduce the shape.

I could count the faces by colouring them as I went.

I could count the edges by colouring them as I went.

This shows the screen part way through this process.

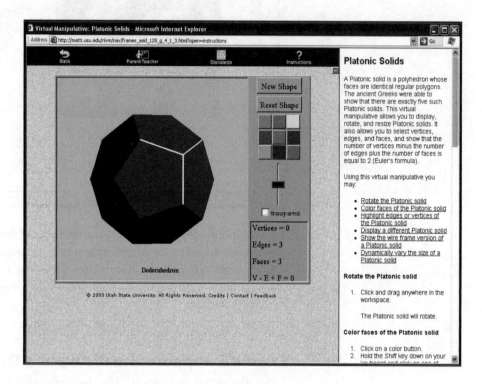

Okay, so this was fun, but in the classroom would it really replace my plastic interlocking polygons and blobs of Blu-tack?

I then looked at another of the files 'Platonic Duals'. In this piece of software, having selected a platonic solid, I could choose to view it as a 'wire frame' to see inside the shape. If I then selected 'Show dual', the dual appeared inside the original shape.

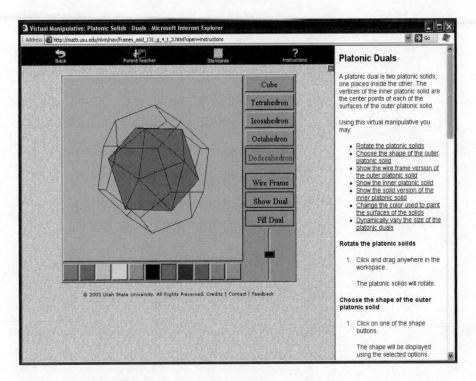

Having created the dual of the dodecahedron, by manipulating the shape on the screen, you can begin to make conjectures about its properties. How many vertices, faces and edges does it have? Why? What shape is it?

This piece of software was clearly offering an image which would enable pupils to explore the 'insides' of shapes in a virtual environment, something which would not be easy to do with traditional tools.

The arrival of the data projector has caused many schools to reconsider how they can make the most of one computer in the classroom. The examples in this chapter illustrate the power of the visual image to develop mental models and visualization skills and bring the outside world into the classroom. It is hoped that, by using such tools, pupils' motivation and engagement in learning mathematics will be enhanced and their learning experiences enriched.

Chapter 4

Mathematics moving with ICT

In this section you will:

■ find out about kinesthetic learners and how using kinesthetic teaching approaches in mathematics can support *all* pupils;

■ consider some kinesthetic ICT-based learning activities that can be incorporated into mathematics lessons;

■ look at some ICT tools that let us manipulate mathematics kinesthetically including interactive whiteboards.

For many of us the kinesthetic aspect of learning mathematics is not obvious. If, like me, you spent most of your secondary school mathematics lessons sitting in rows with a textbook relying on mainly auditory input, it may be a completely unfamiliar notion!

To find out more about the thinking behind kinesthetic learning, we begin with Howard Gardner, who, in 1979 in his book *Frames of Mind*, proposed a theory that people possess 'multiple intelligences'. This was a redefinition of all aspects of intelligence that can be developed through education.

Based on his extensive research, Howard Gardner first concluded that there are seven intelligences; linguistic, logical-mathematical, musical, bodily-kinesthetic, spatial, interpersonal and intrapersonal. He later added two more; existential intelligence and naturalistic intelligence. An implication of his theory is that people can be profiled according to how they prefer to learn.

Intelligence Health Warning

The most common misinterpretation of Howard Gardner's work is that pupils can be grouped by learning styles and should be given tasks or activities *only* in a way that is in sympathy with the appropriate style. This completely misses the point! Howard Gardner warns against this on page 89 of his later book *Intelligence Reframed*.

Myth

There is a single 'approved' educational approach based on multiple intelligences (MI) theory.

Reality

MI theory is not an educational prescription. There is usually a gulf between scientific claims about how the mind works and actual classroom practices. Educators are in the best position to determine whether and to what extent MI theory should guide their practice.

Howard Gardner describes some superficial applications of MI theory that he has seen in many hours of observations in schools:

➡ attempting to teach all concepts of subjects using all the intelligences;

➡ believing that going through certain motions activates or exercises certain intelligences;

➡ using intelligences as mnemonic devices, such as remembering a sequence of numbers by singing it!;

➡ conflating intelligences with other desired outcomes, for example claiming to develop inter-personal intelligence through co-operative learning activities;

➡ labelling people in terms of their intelligences.

The bridge between aspects of Howard Gardner's MI theory and classroom practice is supported by a series of books by Alistair Smith and colleagues also published by Network Educational press.

In *Accelerated Learning in Practice* Alistair Smith gives this description of kinesthetic learners.

Kinesthetic learners:

➡ will remember events and moments readily and will also recall their associated feelings or physical sensations;

➡ will enjoy and benefit from physical activity; modelling; body sculpture; field trips; visits; learning by 'doing';

➡ may spell best when able to replicate the physical pattern of the letters of the words either by writing or by moving the writing hand or by rehearsing such movement as the letters are spelled out;

➡ may be characterized by use of accompanying hand and body gestures while talking but not to reinforce meaning; often it will be a physical and repetitive patterning of small movements as one talks or listens;

➡ will fidget and need regular breaks when learning;

> ➡ will give instructions by demonstration or modelling with the body or with gestures; when giving directions would be more inclined to take you there;
>
> ➡ will use kinesthetic predicates such as 'It feels good to me', 'I don't follow'.

Some of these can be interpreted more specifically within the mathematics classroom.

Some kinesthetic approaches for mathematics

Exploring place value and multiplying and dividing by powers of 10

A set of chairs are laid out in a row with the labels 10 000, 1000, 100, 10, 1, 0.1 and so on, as appropriate for the class. Pupils are given digit cards and asked to sit on chairs so that a specific number is represented. A prominent object representing the decimal point is 'nailed' between the 1 and the 0.1. (Pupils that I taught have not forgotten seeing me hammering a nail into the wall as a physical decimal point, and associating my actions with the important mathematical fact that the decimal point never moves; it is the digits that move!)

Pupils are then asked to move so that they represent the result if 356.8 is multiplied by 100 (or divided by 100). The discussion about whether additional zeros may be needed, or have become redundant, is made more poignant when pupils holding zeros move in to sit on chairs or get up from chairs and leave the scene, as the case may be.

Exploring polygons and their angles

Pupils 'walk' the outlines of regular polygons, turning through the exterior angles. As they experience their own movement through a 'full turn' they may relate it to the fact that the sum of the exterior angles of a polygon is 360°. This could be a pupil activity prior to doing some work using LOGO.

Exploring properties of triangles

Pupils explore the accurate construction of triangles with side lengths determined by pieces of string. A group of three pupils selects three pieces of string, each labelled with its respective length, stretch out each piece of string and explore which triangles can, and which cannot, be constructed. As each pupil holds two string ends the three pupils have effectively become a 'dynamic' triangle; they can change their relative positions to try to find a triangle that works.

The lengths of string available determine the degree of challenge of the activity. For example, the set of lengths 1 metre, 1.5 metres, 1.8 metres, 1.8 metres, 3 metres, 4 metres and 5 metres would allow them to explore the fact that the sum of the two shorter sides of any triangle must be greater than the length of the longest side.

People loci

Pupils learn about geometrical construction techniques through 'people loci' activities. For example, mark a fixed point on the classroom floor and ask pupils to move so that they are always a fixed distance from the point.

What shape do they trace? (They can leave a trail of 'Post-it' notes or multilink cubes behind them to trace their path.)

What if they move so that they are always a fixed distance from a wall?

What if they move so that …?

Using data-loggers to draw 'live' distance-time graphs

Try giving the problem below to some pupils. Listen to their discussions without intervening.

Laura and Nadim are discussing the following distance-time graph.

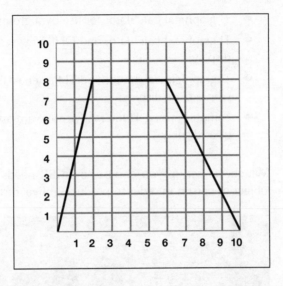

I think it is showing someone walking to the top of a hill and across the top and then down again.

No, I don't think we can tell where they are. I think they are moving, and then standing still for a bit and then moving again.

Who do you think is correct?

Did your pupils resolve the conflict?

How did they do this? ... or did they not?

What difficulties did they encounter?

Did they comment on the lack of units on the axes?

What mathematical vocabulary did they use?

Clearly, it would help pupils to interpret distance-time graphs if they were involved in creating them! A very simple piece of ICT kit is a data-logger that detects motion. For example the Calculator Based Ranger (CBR®) from Texas Instruments (TI) has been on the market for about 10 years, and many descriptions and discussions of its use in the classroom have been published. In the NCET publication *Data Capture and Modelling in Mathematics and Science* one of the pupils describes his experience:

'Thinking about the lesson, having made someone walk up and down the classroom made it easier to understand the graph and decode the information that was given to you.'

The CBR can be used in several ways in the classroom, giving you lots of flexibility as to how you plan the lesson. It can be connected to a:

➡ TI graphical calculator for use by a group of pupils;

➡ TI graphical calculator and OHP Viewscreen panel for whole-class display;

➡ TI graphical calculator and TI Presenter through a data projector for whole-class display;

➡ computer using TI Interactive software and a data projector for whole-class display.

Whichever set-up you choose, the CBR needs to be pointing at the moving object or person in order to produce a 'live' distance-time graph.

Once the CBR is running, the graph is plotted on the display screen as the movement takes place. This has a huge impact on pupils' understanding of the concept. They are able to simultaneously connect the visual experience of seeing the movement with the resulting graph.

What might pupils learn in this environment by asking effective questions such as 'What is it possible to find out from a distance-time graph?' and 'What is it not possible to find out?'

Some possibilities are:

– that a horizontal line 'happens' when there is no movement;

– that faster movement produces a steeper graph;

– that you cannot tell if someone is walking backwards or forwards, or waving or smiling or …;

– that you don't know where the movement took place; in a classroom, on a hill...;

– that you don't know whether the movement was walking or running or cycling or ..., although some calculations and estimations may help.

Many teachers have seen the CBR in action on professional development courses, but not many schools have integrated its use into lessons. This may be due partly to the fact that you need to invest development time in planning some activities, preparing associated resources and amending schemes of work.

A range of classroom resources have been produced to support you to experiment with using a CBR in the mathematics classroom.

WEBSITE

➡ There are some such resources within the 'Enhancing Subject Teaching Using ICT' program. Visit www.cpd4maths.co.uk for more information about these.

➡ The Mathematical Association has produced an exemplar lesson that uses the CBR, which is freely available from the Becta website. This lesson is one of the video case studies on the CD-ROM produced by the DfES ICT in Schools team. It features Maria Stewart using the CBR with a Year 8 class at Davison High School for Girls, Worthing. Maria's lesson plan is in Appendix 3(a).

➡ Most LEA mathematics advisers and consultants have access to a set of TI calculators and CBR for loan to schools; you may be able to borrow the kit!

In my own teaching experience, I have used the CBR with groups of pupils from Year 4 to Year 13, and they have all thoroughly enjoyed the experience. They were able to self-assess exactly what they learned during a lesson, and say how the kinesthetic approach was a significant factor.

In each of these examples, what is important is that the pupils' kinesthetic involvement in the lesson has supported them to develop in their minds a key mathematical idea or concept. It may be that the pupils who enjoy learning kinesthetically readily volunteer to be the 'movers', while other pupils may prefer to observe or record; what matters is that *all* pupils experience the activities in some way.

What about a different approach to kinesthetic learning in mathematics?

So far we have looked at how *moving themselves* supports pupils to learn mathematics. However, another dimension opens up if we include the notion of the pupils *moving the mathematics*!

ICT is bringing a new kind of kinesthetic learning into the mathematics classroom. Whereas previously it was the pupils who did the moving, with some ICT resources it is possible for the pupils to move the mathematics. We have already seen this in action in Chapter 1 with the use of dynamic geometry software to explore transformation. By actually transforming mathematical models, pupils are able to experience mathematics in a new way.

In a traditional classroom, this can happen when pupils use practical resources such as Multilink Cubes, number lines, protractors and compasses as the means for exploring mathematical ideas through problem-solving activities.

However, giving pupils the resources does not imply that they will automatically move the mathematics!

Consider the following two activities, which address similar themes:

How many Multilink Cubes do you need to make a 2 x 3 x 4 cuboid?	Take 24 Multilink Cubes. How many different cuboids can you make?

The first activity is a closed task in which the pupils may use the Multilink to construct the model.

The second activity may require the construction and reconstruction of a range of models, and also supports pupils to look at cuboids from different perspectives. The Multilink Cubes would be an essential tool for pupils when solving the task and convincing themselves that they had found all of the solutions.

The remainder of this chapter will focus on the range of ICT tools available to support pupils as they interact dynamically with mathematics. As in previous chapters, dynamic mathematics needs to be explored dynamically, not statically by reading a book, so load the CD-ROM and read on.

One of my own memorable mathematics moments was possibly one of the main influences that inspired me to begin using ICT in my teaching. I was taking part in a mathematics professional development course, with the course tutor Warwick Evans. We began with a practical task, which I will describe.

Take a circle of paper, about 15cm in diameter. A piece of filter paper from the science resources works well!

Mark a dot on the paper anywhere other than at the centre.

Fold the edge of the circle so that the circumference just touches your mark. Make a firm crease.

Unfold the paper.

Fold the paper from a different place, again so that the circumference just touches your mark.

Keep doing this so that you produce an 'envelope' of folds on your circle of paper.

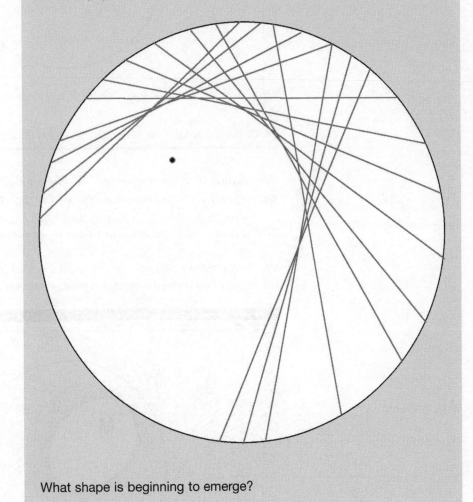

What shape is beginning to emerge?

We then turned to the computers and loaded Cabri-Géomètre (which in 1995 was a 'new' genre of computer software called dynamic geometry software!). We learned how to draw a circle...

... and construct a point inside it.

We then became engaged in an electronic version of the paper-folding task that we had just completed practically. A discussion began about whether the fold lines were each a perpendicular bisector of the line joining a point on the circumference and the marked point.

A few minutes later we had the following picture.

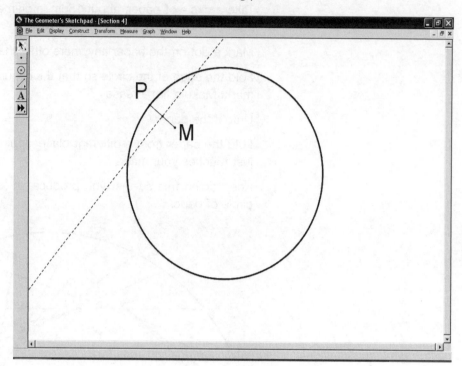

We explored what happened if we moved the point P around the circumference of the circle and noticed the path that the dotted line took.

This was my first experience of dynamic geometry!

We then used the software to show all the fold lines at once, by constructing the locus of the dotted line as the point P moved around the circumference.

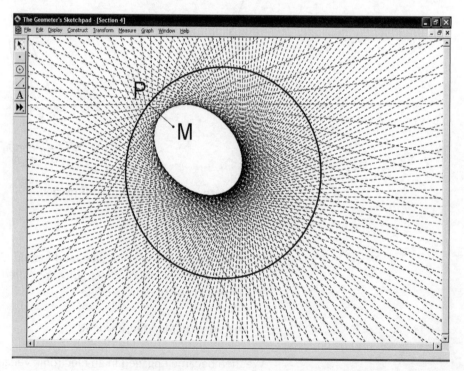

This was amazing enough! But then we explored what happened if we moved the point M; something that we couldn't do with the paper circle without starting again. What if it was near the centre? ... or at the centre? ... or near the circumference ... or outside the circle altogether...?

CD-ROM

I am not going to give away what happened next. Open the dynamic geometry software file (*Chapter 4 circle.gsp*) on the accompanying CD-ROM and try this for yourself!

At that moment I knew, from my own experience, that ICT was going to change the way that I taught mathematics and the way my pupils would learn mathematics. I also realized that I had become a learner again; with this tool I could begin to explore geometry in a way that was never possible when I was at school.

In 2001 The Royal Society published a report *The Teaching and Learning of Geometry 11–19* which said:

'Our view is that teachers should now have at their disposal an appropriate variety of equipment from which to select, depending on fitness for purpose. In particular we wish to see the potential of ICT realized in supporting the teaching and learning of geometry. There is already software available, such as for dynamic geometry (DGS), but its use is not widespread. Many schools do not have licences for the software. There is also a need for the development of additional software, such as to support work in 3-dimensions. Increasing numbers of schools and colleges are now being equipped with interactive whiteboards – where a computer image is projected onto a touch sensitive screen. This medium has considerable potential for interactive whole-class teaching of geometry.'

Part of the fascination of developing the way you use ICT in your teaching is derived from the evolution of the ICT tools themselves, to which new features are continually added.

While dynamic geometry software began life as an ICT tool within which people could explore shape, space and measures, some packages, The Geometer's Sketchpad and Cabri Geometry, now include a co-ordinate system and function graphing tools that allow you to explore co-ordinates and graphs dynamically. In Chapter 5, we will look at some rich examples of the way software can support us in connecting mathematical ideas across mathematical topics.

When thinking how to use ICT resources to enhance mathematical learning, it is often productive to focus on a topic that is difficult to teach or that pupils find difficult to learn. By doing a little research, you can uncover some of the misconceptions and errors that pupils experience, and consider whether you could adopt a better approach using ICT.

Children's Understanding of Mathematics: 11–16 by Kath Hart and the Concepts in Secondary Mathematics and Science team, describes methods used by children to solve a range of mathematical problems, summarizes their errors and suggests some teaching approaches. This is an excellent starting point if you are looking to develop your own teaching approaches using ICT as a resource.

As an example we could focus on algebra. This is a big topic, which encompasses vocabulary, notation, underlying concepts, graphical representations and images. Ideas that pupils may find difficult include:

- understanding that in $3a + 7 = 20$, a has a unique value, of $4\frac{1}{3}$, whereas in $3a + 2n = 20$, a can have an infinite number of values;

- understanding that $3a + 2n = 20$ is inherently the same as $2x + 3y = 20$ or $y = 6\frac{2}{3} - \frac{2}{3}x$ on this graph.

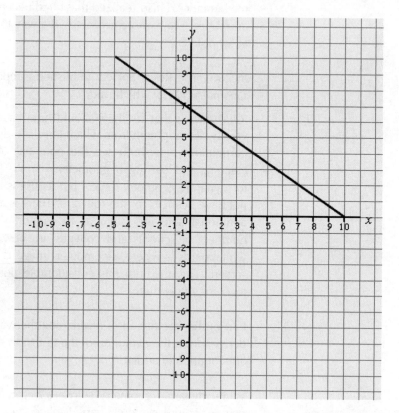

How might we develop a mathematical image that will give pupils a picture to hold in their heads that will support their emerging understanding of algebra? It may be possible to build in something dynamic, which will allow us to use one of the most effective questions in mathematical inquiry:

'What is changing and what is staying the same?'

or

'What is variable and what is constant?'

The following example grew from an idea by Adrian Oldknow. He proposed a dynamic number line, constructed in dynamic geometry software, on which a variable point, *n*, is plotted. Other points are plotted on the line in various positions in relation to *n*.

As the point *n* is moved on the line, so the other points also move, while retaining their original positions in relation to *n*.

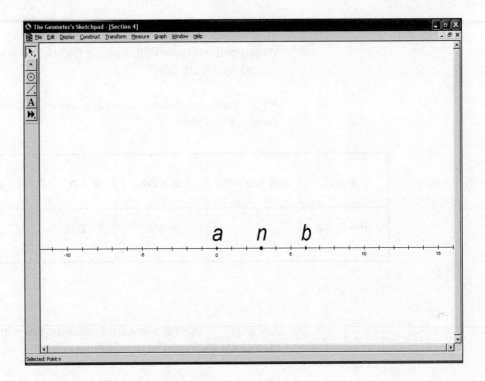

If you were to show the number line above to pupils, your questioning might go something like this:

Look at the number line.

What are the values of *a*, *n* and *b*? (*a* = 0, *n* = 3 and *b* = 6)

Can you suggest any relationships between the values of *a*, *n* and *b*?

(n is 3 more than *a*, *b* is 3 more than *n*, and so on)

In the classroom you might record pupils' suggestions on a flip chart or whiteboard.

You might then invite a pupil to move *n* to a new position on the line, for example to where *n* = 1.

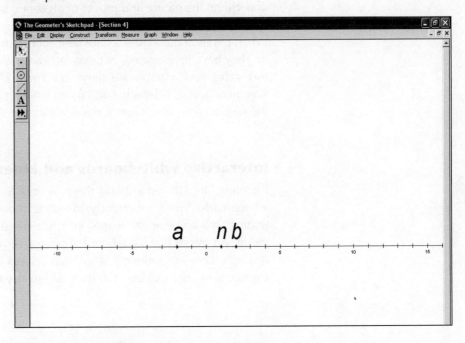

You could now ask:

Which of the relationships that we had before are still true?
... and which are false?

An alternative pedagogic approach to this task might be to give pupils a set of cards such as these:

$a = 3$	$a + b = 6$	$b = 2n$	$a - b = 10$	$a + 3 = n$	$n = 10$
$n - 3 = a$		$n = \dfrac{a}{2}$	$b = 7$		$n = \dfrac{b}{2}$

Ask pupils to use the software to explore the number line for themselves and then place each card in one of three columns:

always true	**sometimes true**	**never true**

Include some blank cards on which pupils can create their own statements. Observing what pupils write on the cards is an additional opportunity to assess pupils' understanding of the underlying concepts.

How does the movement of variables on the dynamic number line help pupils understand the nature and power of algebraic statements?

When pupils come across a statement such as $n = a + 3$, how do they interpret it? They have investigated, both visually and dynamically, the relationship that this states. As a result are they now more likely to appreciate that, if the statement is true, although a and b can have an infinite number of positions on the number line, the value of n will always be 3 more than the value of a.

Interactive whiteboards and kinesthetic approaches

Presently, in UK education, there is much excitement about interactive whiteboards. We are currently investing more per head of population on putting interactive whiteboards into schools than any other country in the world. An interactive whiteboard appears overnight in the classroom of many teachers. They are suddenly faced with a steep learning curve as they begin to explore how they can use it in their day-to-day teaching.

Initially it may seem a slightly risky business to expose your own ICT strengths and weaknesses in full view of the pupils. But, once the technology is switched on, the question arises: 'How will the interactive whiteboard enhance the pupils' mathematical learning experience?' When we think about how the software is going to be driven from the front of the classroom, a kinesthetic link emerges.

Many teachers immediately grasp PowerPoint as a convenient piece of software. Into it they can put text, graphics, sounds and animations to create a set of slides, which constitute the resources for the lesson. However, beware of justifying the expense of an interactive whiteboard if the only time it is touched is when you tap the board to advance to the next slide!

In Chapter 5 we will look in more detail at some of the advantages and disadvantages of a PowerPoint-style approach. But we will explore alternative presentation tools, some of which are particularly powerful when used in conjunction with an interactive whiteboard.

In Chapter 2, we saw an example of Research Machines' 'Easiteach Maths'. This is essentially a presentation tool. It allows you to prepare slides that include a range of mathematical images, such as number lines, grids, polygons and graphs. You can also work with pre-written files such as those that are provided within the RM MathsAlive Key Stage 3 scheme. The software itself is not mathematically intelligent; for example if you type 3 + 6 = 10, there will not be any feedback to tell you that you are wrong!

But, as the screenshot below shows, high quality, accurate diagrams can be produced in minutes, and all these images – the number lines, arrows, digit cards and fraction pictures – can be dragged, manipulated and coloured.

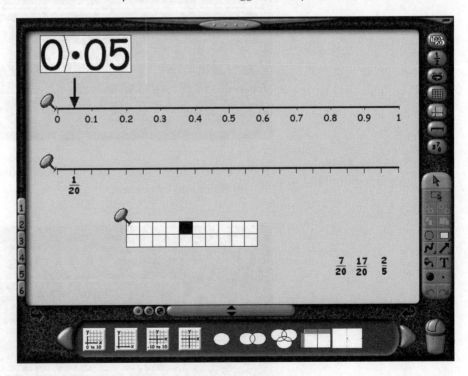

The focus of the activity then becomes the questioning!

How could you shade $\frac{7}{20}$ of the rectangle?
Where is $\frac{7}{20}$ on the decimal number line?
What about $\frac{2}{5}$?

This is a good example of the importance of the teacher's role; the ICT could never replace you as the teacher. If you look at the screen, there is nothing to tell the pupils what to do! Instead it makes you the most valuable asset in the classroom. Given this ICT tool, how can you create teaching resources that will support you to question pupils' understanding, actively address pupils' misconceptions or help pupils construct new mathematical meanings?

Previously, in Chapter 3, we looked at 'Division Exploded' from John McCormack. The complementary software 'Multiplication Exploded', in which pupils explore multiplication, offers a kinesthetic learning environment with some very dynamic features. A 30-day trial version of this software can also be found on the CD-ROM.

CD-ROM

The software is designed to help pupils *understand* the concept of multiplication by allowing them to construct a calculation, and then explore the grid method before developing a traditional vertical algorithm.

The opening screen looks somewhat 'scary'. However, the word 'grid' is a clue that all is not what it seems!

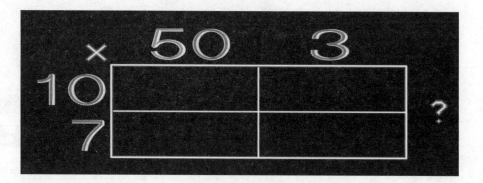

At this point, if pupils know which numbers to place in the grid, they can type them in and get feedback indicating whether they are correct.

But selecting the '?' moves you to a very exciting feature which reveals the power of this software.

A hammer is revealed. When you 'hit' a number with it, it breaks the number into its factors. For example, hitting the 350 produces **5 × 10 × 7**. Pupils might want to rearrange this product to give **5 × 7 × 10** enabling them to see more easily that its value is **350**.

Clicking a multiplication sign recombines the numbers.

The numbers can be broken down fully to reveal how the prime numbers are 'building blocks' of our number system – a separate exploration in itself!

The next stage in the progression is to link the grid method of multiplication to the traditional algorithm. As the grid is filled, so the software requires pupils to drag the numbers from the grid, revealing the next representation.

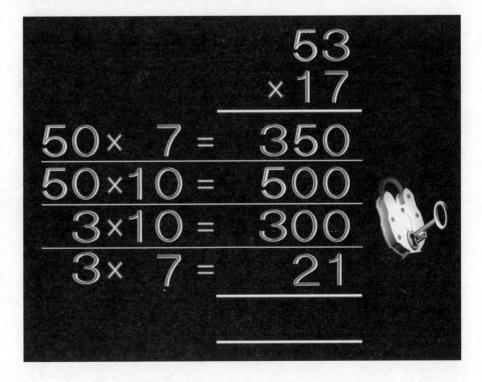

The padlock icon allows for two routes at this point. In its unlocked state, pupils can drag products and combine them to confirm mental additions.

The screen below shows the **350** and **500** combined to give **850**. It also confirms that the number **850** is the value of **50 × 17**.

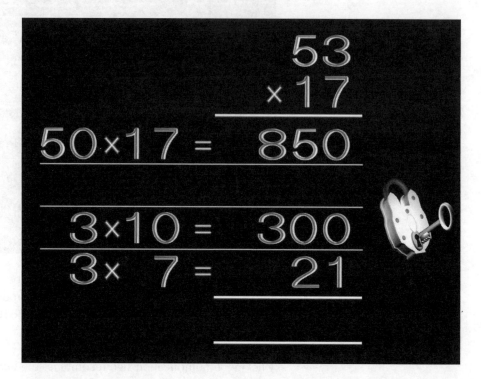

If the padlock is locked, pupils are moved towards a vertical addition strategy.

Used on the interactive whiteboard, this software gives pupils the feeling that they are physically interacting with numbers – pulling them apart to partition them and dragging them together to recombine them. The hammer image is powerful in that pupils actually see numbers being 'broken' or 'split' into factors. 'Splitting into factors' is a metaphor that teachers often use when trying to encourage pupils to develop helpful mental models or images.

The final example in this chapter was developed to support the plenary phase of an activity which focuses on the following problem:

> Using squares of side lengths 2, 5, 7, 9, 16, 25, 28, 33 and 36, can they be arranged to fit inside a rectangle of size 61 x 69?
>
> 4229 into 4209
> If so how? No
>
> Can they be arranged in more than one way?

Have a go at the problem for yourself.

Choose whichever resources will help you to solve it – squared paper, scissors, Multilink Cubes, rulers, and so on.

Given a free choice of resources, the tools that you choose affect both the way you approach a problem and, in some cases, the conclusions that you come to. To encourage teachers to think about this, I have many times challenged teachers on professional development courses with this problem. From my observations of people in this situation I have identified the following types:

The 'Sceptics' begin by thinking it is a trick question, and immediately check that the sum of all the individual areas of the squares equals the area of the rectangle! These people often only approach the problem numerically, and draw just one diagram at the end to show a correct arrangement. They are not put off by their own inaccuracies, mentally ignoring gaps and overlaps.

The 'Talkers' immediately find each other, and begin to collaborate beautifully, sharing the workload and discussing co-operatively to arrive at mutually agreed outcomes. They tend to be tolerant of others' choices of resources and often say at the end 'If I had been working on my own I would have approached it differently'!

The 'Doers' nearly always go straight for the paper and scissors, and proceed to cut out the individual squares. This group will appear to not be thinking about the solution until they have all of the pieces cut out in front of them. Interestingly, those who write the areas on the squares adopt a more mathematical approach than that of those who don't. By losing sight of the numbers, the latter group will often tackle the task merely as a jigsaw puzzle, not considering other mathematical aspects until they have completed the task, or until questioned!

Of course, with all this going on in the room, it is impossible for groups to remain uninfluenced by what others are doing. This leads to a period of unease as individuals look around the room and begin to doubt their own approaches. Most say that they feel threatened by the busy appearance of the 'Sceptics'. The 'Sceptics', in turn, are oblivious to what others are doing, and say afterwards that they couldn't be bothered with doing any cutting out! At this point it is tempting to wonder how the different approaches may relate to the individual teachers' preferred learning styles.

The most interesting phase within this activity is the discussion during the plenary session. Those who have taken a practical approach find it so much easier to describe their outcomes to others. Having the actual pieces enables the whole group to talk about how they conjectured a solution and proceeded with the task.

On one occasion Nicola Pruden, a Key Stage 3 consultant in Essex LEA, transferred the problem to dynamic geometry software to support the plenary discussion.

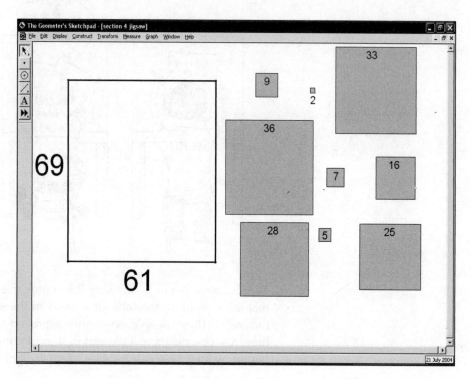

The movable squares can be dragged inside the rectangular frame to represent the task. (Many teachers would have used an overhead projector in a similar way.)

The plenary was substantially enhanced by the ease with which the interactive whiteboard was used with the software, as people talked through conjectures that they had made and explained subsequent approaches that they had adopted.

Far from being merely a 'summing up', the discussions moved towards why a particular approach had been taken, why other approaches had been rejected and why most people were convinced that, allowing for reflections and rotations, there was only 'one' solution.

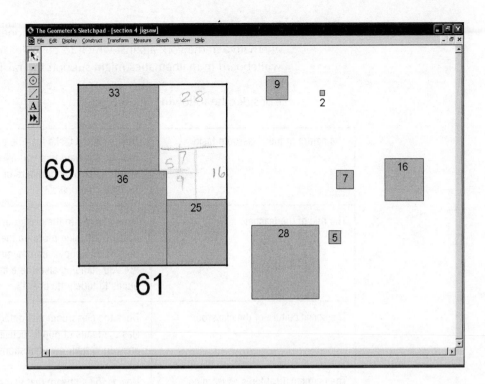

For many people the starting point was to look at the sums of various combinations of the numbers, leading to the conjecture that the square of area 36 had to sit in one of the corners.

In this example, the tactile nature of the ICT resource supported the mathematics.

How can we maximize mathematical learning using interactive whiteboards?

With the current high levels of investment in schools' ICT infrastructure, many schools will have to justify their expenditure on interactive whiteboards on the basis of an improvement in pupils' achievement. The use of interactive whiteboards is still very much in the early stages of development and it is still too early to say just *how* their use is affecting pupils' learning.

A group of primary teachers from Hampshire LEA have been working together on just this issue for a year, and there is a lot of emerging good practice. However, it is proving challenging to say exactly what impact the use of the interactive board is having on pupils' mathematical understanding. This book is full of reasons as to how and why ICT can enhance mathematical understanding. But it is harder to pin down the advantages that an interactive whiteboard brings over and above the use of just a data projector.

The evaluation tool that was provided in Chapter 1 might prove useful here. Think of a lesson or activity in which the interactive whiteboard has been an integral part of the task, and carry out the evaluation activity. Which of the mathematical processes were directly supported by the use of the interactive board?

The 'five dimensions of the classroom', mentioned in Chapter 2, provides an alternative framework against which to consider *how* the use of an interactive whiteboard in mathematics might support the raising of achievement.

Consider the following questions:

The nature of the classroom tasks	How have the tasks that the pupils are engaging in changed? Has the software that you have used enabled you to move sliders to change numbers or drag mathematical images *by touching the board*?
The role of the teacher	Has your role in the classroom changed? Are you focusing more on the questions that you ask in response to pupils' comments and actions? Are you standing elsewhere in the room and allowing the pupils to touch the board?
The social culture of the classroom	Does the classroom feel different? Has the focus of pupils' activity become asking and answering their own questions?
The mathematical tools as learning supports	How is the software that you are using on the interactive whiteboard supporting the learning? Is it dynamic – are pupils being forced to engage in real mathematical activity rather than just operating software?
Equality and accessibility	Has the interactive whiteboard supported *all* pupils to become engaged in learning? Has the board become everybody's mathematical test bed?

As the teacher, you have the huge responsibility of deciding how you are going to use the interactive board in your teaching. This means that you have to be discerning when you are choosing tasks and activities.

Afzal Ahmed and Honor Williams, in their series *Numeracy Activities: Plenary, practical and problem solving* suggest the following classroom strategies for developing talking, listening, reading and writing in mathematics:

➡ Involve pupils in simple starting points and then try asking them how they might vary these or what questions they could think up to answer next. Collect together pupils' suggestions for variations or questions, perhaps on the board or on a large sheet of paper, and try inviting them to follow up a suggestion of their choice.

➡ Ask pupils to keep a record of questions or other ideas they have not attempted. Encourage them to choose one of these questions to tackle on appropriate occasions.

➡ Display examples of pupils' own questions. Invite groups to look at and perhaps work on other groups' questions.

➡ Do not always give pupils rules that work; invite them to try some that do not work and say why they do not.

➡ Encourage pupils to find methods and rules for themselves. Try to involve pupils in comparing the methods in order to agree on the most efficient. See if you can think of ways of involving pupils in generalizing for themselves.

➡ When you want pupils to practise skills, think whether it would be possible for such practice to emerge through pupils' own enquiries or problems which will necessitate the use of these skills.

➡ Think how you might 'twist' tasks and questions in textbooks, worksheets and test papers so that pupils can become more involved in making decisions, describing patterns and relationships and testing conjectures.

➡ Help pupils to appreciate the importance of asking questions such as. 'Is this sensible?', 'Can I check this for myself?'. Offer activities which involve pupils in decisions relating to the 'correctness' of a piece of mathematics.

➡ Show pupils examples of mistakes. Ask them to sort out what the mistakes are and to think how they might have arisen.

➡ Consider how you might incorporate the terms and notations which you want pupils to learn, so that meaning can be readily ascribed to them and so that they can be seen as helpful and necessary.

In all these suggestions, a common factor is that the pupils are being encouraged to do the thinking. They are not just responding to your questions, they are being encouraged to ask their own. One of the great advantages that an interactive whiteboard seems to offer is that it can operate as a big mathematical test bed that everyone can see and interact with. Used in this way, we are a million miles away from tapping the board to advance a PowerPoint presentation! Many of the ICT-based activities in this book, particularly those involving the use of dynamic geometry software, are further enhanced when used with an interactive board.

Chapter 5

Connecting ideas in mathematics with ICT

In this section you will:

■ learn about successful approaches used in mathematics classrooms by teachers described as 'connectionist';

■ find out about concept mapping techniques and how they might support teaching and learning in mathematics;

■ look at how ICT can support the explicit connection of mathematical ideas.

▌Why would we want to connect mathematical ideas?

In 1997 Mike Askew and a team based at King's College, London published an important research report *Effective Teachers of Numeracy*. The team had studied 90 primary teachers and their pupils with the aim of identifying aspects of the teachers' beliefs and practices alongside the progress that their pupils made in mathematics.

The research team found that the following approaches seemed to contribute to making mathematics teaching effective:

➡ stressing the connectedness of numeracy ideas rather than compartmentalizing them;

➡ using pupils' descriptions of their own methods and reasoning as starting points for engaging with numeracy concepts;

➡ placing emphasis on enabling pupils to select strategies according to whether they were both effective and efficient;

➡ emphasizing the development of mental skills;

➡ ensuring that all pupils are challenged;

➡ encouraging purposeful discussion about choice of strategies;

➡ using assessment to inform planning and teaching.

The research team classified the most effective group of teachers as 'connectionist', and the research described some of their common classroom practices as follows.

'Those with a connectionist orientation encouraged methods that placed the highest priority on mental methods. They also regarded it as important that pupils were aware of different methods of calculation and were able to choose methods in relation to their effectiveness and efficiency in solving the problem.

Teachers with a connectionist orientation emphasized the complementary nature of teaching and learning and valued classroom activity, which involved pupils working together with other pupils and teachers to overcome difficulties and to reach shared understandings.

Many teachers with a connectionist approach believed that mathematics should not be taught in a fragmented way and that, where appropriate, pupils should be introduced to some of the complexities of mathematics.

Connectionist teachers who believed that the major factor in learning is that pupils engage and struggle with processes, used pupils' errors as a means of engaging with them in order to further their understanding.'

WEBSITE

An excellent summary of the research and some of the case studies within it is available on the General Teaching Council website at http://www.gtce.org.uk/research/numeracyhome.asp.

We are immediately faced with a dilemma!

Most school mathematics curricula can't help but compartmentalize mathematics!

The English and Welsh national curriculum compartmentalizes the curriculum into four strands: Using and applying mathematics; Number and algebra; Shape, space and measures; and Handling data.

The Primary and Key Stage 3 Frameworks for teaching mathematics sub-divide the curriculum even further with topics such as Sequences, functions and graphs; Transformations; and Processing and representing data!

However, look out for the bold blue print at the bottom of the pages in the respective supplements of examples – some of them have explicit 'Link to…' statements, highlighting where connections across topics and strands can be made. We need to keep in mind when we are teaching how to make the connections explicit and encourage pupils to make connections for themselves.

▌Connecting ideas using concept maps

The book *MapWise: Accelerated learning through visible thinking* by Oliver Caviglioli and Ian Harris, also published by Network Educational Press, gives much of the background to, and examples of, the way that concept mapping techniques enhance learning.

In their book they describe how 'model mapping':

➡ infuses thinking skills into subject delivery;

➡ supports each stage of the accelerated learning process;

➡ can be used to measure and develop intelligence;

➡ is the essential skill for including and supporting children with dyslexia;

➡ supports pupils of all learning styles in developing their essential learning skills;

➡ supports teacher explanation and pupil understanding;

➡ makes teacher planning, teaching and reviewing easier and more effective.

There is a range of vocabulary associated with mapping activities – model maps, mind maps, concept maps, splurge diagrams, spider diagrams, thought showers, all of which may have subtle similarities and differences. However, a common feature of them all is that the thinking behind their creation is unique to their owner or owners. The process of creating or reviewing a map is at the heart of the learning activity.

Here is a simple example:

As part of the National Numeracy Strategy training courses offered to primary and secondary teachers a video of a lesson by a Year 6 teacher, Emma, shows her using a mapping strategy to enable her Year 6 pupils to review their understanding of finding different percentages of a given amount.

Pupils were given a large sheet of a paper, on which was written a statement similar to this one.

$$\text{360 is 100\%}$$

From this initial statement the pupils were asked to deduce other correct mathematical statements, and show them on the sheet of paper.

This led to responses such as:

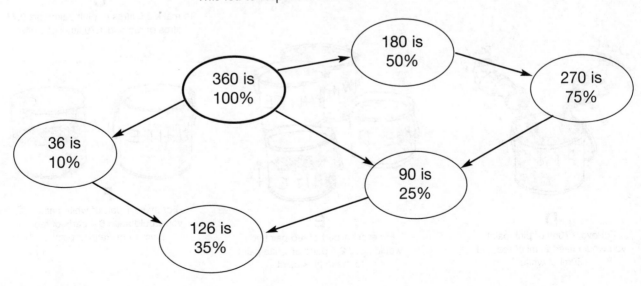

The video of the lesson shows the pupils discussing how they are calculating the percentages mentally and combining previous answers to create new ones.

Alternatively, the task could also have been given in a linear fashion, with the teacher writing the following on the board:

360 is 100 per cent

and generating
180 is 50 per cent
90 is 25 per cent

However, would a linear approach have supported the pupils to think creatively about different *combinations of numbers?* A subtle difference in the presentation of the task has placed an emphasis on the mathematical processes since, in making an actual connecting line on the diagram, the teacher can ask pupils 'How have you connected these two pieces of information?'.

In his book *Active Whole-class Teaching*, Robert Powell describes some concept mapping activities. One example requires pupils to 'display a visual representation of understanding' by asking pupils to connect two ideas and explain how the two are connected. His example relates to the causes of the collapse of the Tsarist regime in Russia in 1917. The approach could easily be adapted for a mathematical theme, say ratio and proportion.

A
To make pink paint, red and white paint is mixed in the ratio 2:5

B
Mix 4 'dips' of red paint with 10 'dips' of white paint to make pink paint

C
To make 2.5 litres of pink paint, mix 0.71 litres of red and 1.79 litres of white

D
For every 70ml of pink paint you would need 20ml of red and 50ml of white

E
For every 1 part of red paint you would need 2.5 parts of white paint to make pink paint

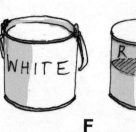

F
For every 1 part of white paint you would need 0.4 parts of red paint to make pink paint

The first thing that you notice is that there are no questions – there are no 'answers' to find! The task for pupils is to decide on two statements that they can connect with a line and then explain the connection.

Pupils could record their discussions by completing a framework such as:

Statement _A_ connects with

Statement _B_

Reason:
If you double the amount of red paint, you need to double the amount of white paint.

Calculation:
2 X 2 = 4
2 X 5 = 10

There is much interest in developing the use of concept mapping techniques in education. Here are a few references to other resources that support their use in mathematics:

> In their book *Adapting and Extending Secondary Mathematics Activities: New tasks for old*, Stephanie Prestage and Pat Perks of the School of Education, University of Birmingham include a chapter called 'A splurge of ideas'. They show how what they call 'splurge diagrams' can be used to generate thinking about a mathematical topic. They suggest that creating a splurge diagram can be a good starting point when planning a teaching sequence.

WEBSITE

> Learning and Teaching Scotland have created an area on their website to support the use of concept mapping techniques for mathematics:
>
> http://www.ltscotland.org.uk/5to14/specialfocus/mathematics/mindmapping.asp

> 'Assessing understanding in mathematics with concept mapping', written by Arthur Baroody and Bobbye Hoffman Bartels, is an excellent article in *Mathematics in School*, the journal published by the Mathematical Association. Arthur and Bobbye give examples of how they have used concept maps to assess the accuracy of the connections that pupils make when connecting ideas with a specific mathematical focus.

So how can ICT support concept mapping?

Concept mapping software was originally aimed at the business sector to support brainstorming and project planning sessions. However, it did not take long for educators to see the potential. There are now several concept mapping software packages on the market which have been designed specifically for education: MindGenius; Kidspiration; SmartIdeas. These all enable you to produce an electronic concept map. They include features which allow some of the thinking behind the map to be recorded, images to be added and hyperlinks to other files to be established.

This is best explained using an example.

The following concept map was produced using SmartIdeas, a concept mapping package which is specifically designed for use with a SmartBoard, which is an interactive whiteboard. This example also features the book *Teaching Secondary Mathematics with ICT* edited by Sue Johnston-Wilder and David Pimm.

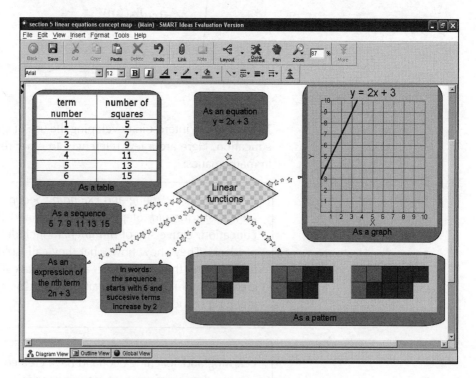

The central idea for this concept map was 'Linear functions', and it was written to support a revision lesson for Year 9 pupils.

On first glance, the map shows seven associated representations of a linear function.

Some initial questions for *you* to consider are:

➡ How many of the ideas can *you* connect?

➡ What mathematical language do *you* need to explain your connections?

➡ Which of the representations would *you* expect pupils of different ages and abilities to be able to explain or connect?

➡ And what language might *they* use or understand?

However, what is missing from this static concept map is the power of the ICT package in which it was created!

In this example, if you double click (or tap if you are at an interactive whiteboard) the 'As a graph' image, the software hyperlinks to a graphing package, where the equation could be altered to something else to follow a new line of enquiry. Or a double click on the 'As a table' image takes you to another file where the values in the table can be altered.

Several pedagogical approaches could be taken:

➡ Alter one of the representations to make it inconsistent and ask pupils to 'spot the error' and suggest corrections.

➡ Only display four or five representations, and ask pupils to add their own ideas.

➡ Ask pupils to connect any two representations and give their reasons.

➡ Using this map as a model, ask pupils to create their own concept map for a different pattern.

In general, concept mapping software has features that will let you:

➡ create a concept map, select colours for branches and rearrange the map dynamically by dragging;

➡ add attachments to the branches, such as image files, audio files and icons;

➡ add hyperlinks to other maps or other software files;

➡ include mental links between branches.

If you think that concept mapping has a place in the mathematics classroom, consider if and how your pedagogy may be changed. For example, you will need to be clear about the purpose of the activity, the context within which you are going to use it and the mapping strategy you will use.

Here are some ideas.

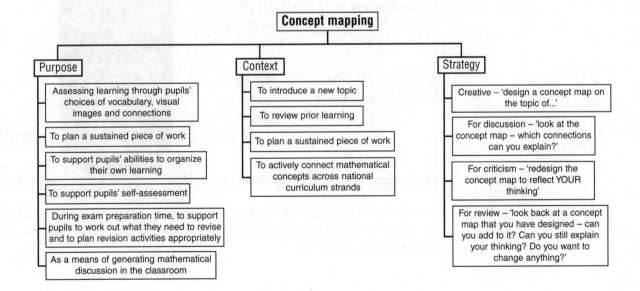

Other approaches that support making connections in mathematics

There are other ICT resources that support teachers and pupils in seeing connections within and between mathematical topics.

An obvious example is the graphical calculator, which has been around for over ten years now, but remains underused in schools. In comparison with computers, their price makes them a desirable option. They can be used in the classroom, giving each pupil access, and they are lightweight and easy to manage.

The facility to display the calculator screen for whole-class teaching has long been available through overhead projector display panels. There are also innovative display methods available using 'virtual' calculator software that displays the whole calculator.

WEBSITE

For some Texas Instruments graphical calculators, which are described as 'Flash enabled', the TI website http://www.ticalc.org/ provides links to free software. This software can be downloaded to give you a replica on-screen calculator that can be used for whole-class teaching.

(Type 'Virtual TI' into the search facility and look for the model of calculator that you are using. Refer to the helpsheet in Appendix 4 for guidance in setting up the software.)

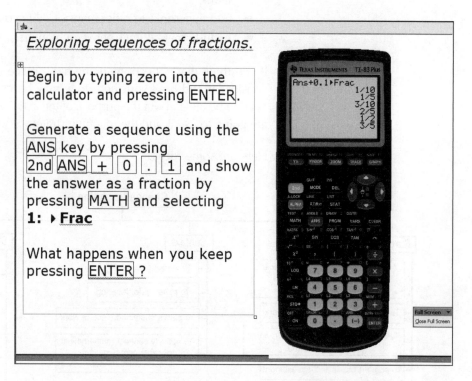

The above activity has been developed from a resource in *30 Calculator Lessons for Key Stage 3*, edited by Barrie Galpin and Alan Graham and published by A+B Books.

By exploring what happens if the 'start numbers' and the 'steps' are altered, pupils can begin to relate their knowledge of equivalent fractions and decimals with that of sequences, making valuable mathematical connections along the way. The example shown is simply a Word document with the Virtual TI calculator running above it. It is a good way to introduce a task that pupils could then explore for themselves using graphical calculators. The plenary phase of the activity, in which pupils share their approaches and findings with the whole class, would also be enhanced.

Connecting algebra and geometry

As mentioned in Chapter 4, dynamic geometry software packages have been developed which allow ideas in algebra and geometry to be explicitly connected.

Let's start with a problem.

Imagine that you have a square of paper or card.

You are going to cut squares from the corners of the paper so that the edges of the paper will fold upwards to make an open tray.

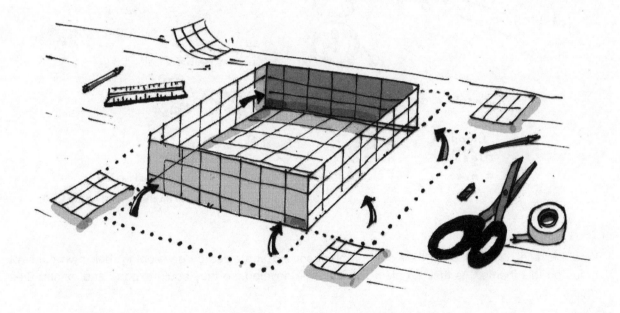

The problem is to work out how big the square cut-outs need to be to produce the largest possible volume for the tray.

Many of you will recognize this task which is often called 'MaxBox' and is a favourite practical task in Key Stages 3 and 4. It is referenced within the Key Stage 3 Framework for teaching mathematics as a Year 9 activity; however, there are many primary pupils who could access this task. I do not believe that rich mathematical tasks should be 'saved' for a particular year group or Key Stage out of fear that pupils might say 'We've done "MaxBox"!' Familiar tasks offer another opportunity to connect pupils' previous knowledge and understanding within a similar context or situation and extend it further. After all, we wouldn't take much notice of pupils who claimed 'We've done fractions'.

Pause for thought…

Before beginning the task, if you were to draw a concept map with the central theme 'problem solving' what would yours look like?
Here is mine!

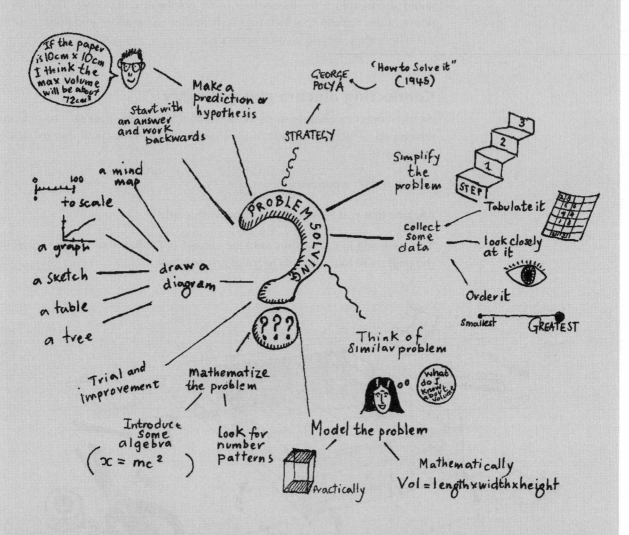

Would it help pupils to build their own problem-solving strategies by developing their own concept maps on this theme? As their problem-solving skills develop, so they could expand and extend their maps.

Back to the task.

Where do *you* go next?

Do you make a model from a square of paper of a particular size?
Do you sketch what it might look like?
Do you dive into some calculations?

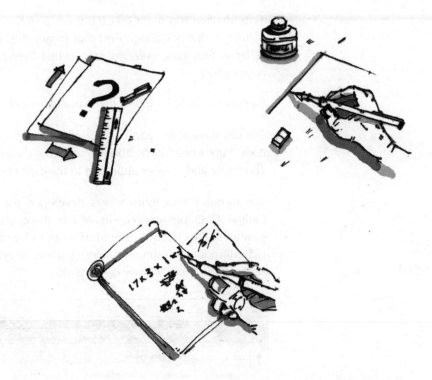

The starting point for everyone is 'Do I understand the problem?' and 'Have I done anything similar before?'

In the classroom, how often do you *tell* the pupils what to do next?

'Okay, I want you to start by taking a 20 cm by 20 cm square of squared paper, cut squares from the corners and work out the volume. Then cut off a bigger corner, collect some results and put them in a table like this....'

How often do you give pupils time to use their own ideas and allow them to discuss and develop systematic approaches, choose their own resources and learn from their mistakes?

Time is often the main constraint on adopting a more open problem-solving culture in the classroom. And that means that you are anxious that pupils get as far as they can, even if it means that they all end up taking a very similar approach!

So how could ICT support this particular task?

The first stage of this particular problem is getting your head around the problem itself. Pupils need to be able to 'see' how the paper folds to give an open tray. The paper and scissors approach to this task is essential for most pupils.

WEBSITE

The resource that follows was developed for The Mathematics Consortium Online CPD programme, in which there are a series of 18 lessons that exemplify the effective use of resources in Key Stage 3 mathematics, including ICT. More information on this programme is available from The Mathematical Association at http://www.m-a.org.uk.

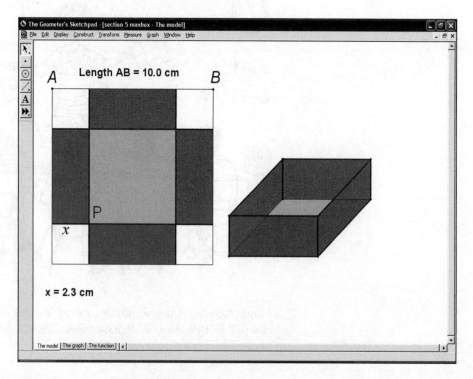

CD–ROM

If you are near a computer, open The Geometer's Sketchpad file *MaxBox.gsp* and select the first page, 'The model'.

The size of the square corners can be varied if you click and drag the point P. The 2-D representation of the box automatically adjusts. This supports pupils to 'see' that you can make a large shallow tray or a tall thin one.

At an appropriate time in the task, pupils may begin to think about graphing the calculated volume of the box against the value, *x*.

Move to the second page 'The graph'.

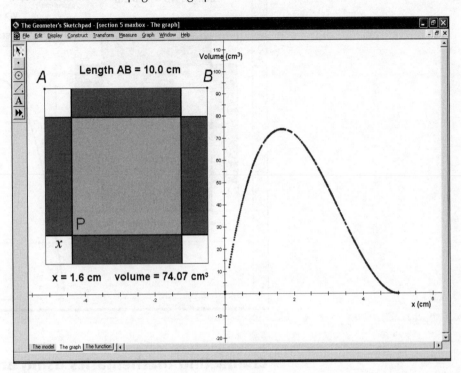

The dynamic nature of the software allows the geometric model to be connected directly to the graph. This time, as the point P is moved, the co-ordinate points representing the corner cut (*x*) and the volume (V) are simultaneously plotted on the graph. If you were using an interactive whiteboard or projecting onto a traditional whiteboard, you might want to ask pupils to predict and sketch the shape of the graph. Again, by using this dynamic image, many pupils will be supported to connect the original practical task with their calculations of the volume and the subsequent graph.

For those pupils who begin to use algebra as a tool in their explanation of their results, the final page 'The function' offers an opportunity to make conjectures about different functions and see how they 'match'. The 'Show function' action button will display the correct function, which can easily be hidden from view while pupils explore their own ideas.

In this example screen view, the superimposed function gives pupils the big picture. That is, there are parts of the function which exist (when x is less than zero or more than 10) but do not apply to the real-life problem. These can become discussion points with pupils, giving them a wider view of the nature of mathematical functions.

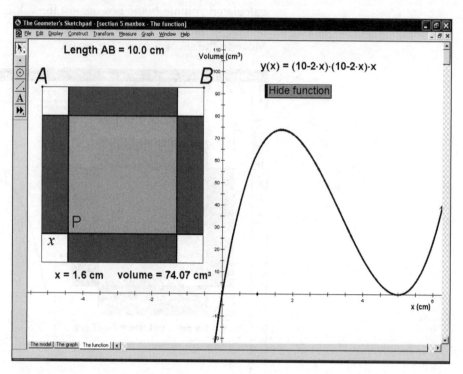

Connecting mathematics using an online mathematics dictionary or thesaurus

Developing pupils' mathematical vocabulary is a key aspect of teaching and learning mathematics. A simple search in Google will reveal a range of online dictionaries with various features. When I searched, I came across a site developed by an Australian teacher Jenny Eather, http://www.teachers.ash.org.au/jeather/maths/dictionary.html. My impression was that it would be a very useful free resource for the mathematics classroom.

WEBSITE

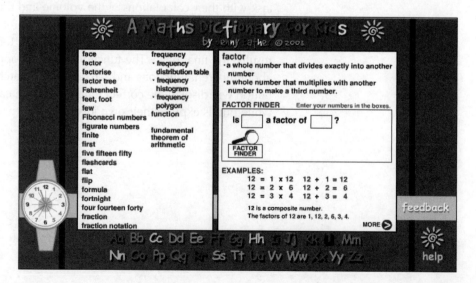

Many of the explanations include short activities, animated images and additional tools. The chosen screenshot shows the word 'factor', and includes a description and a 'Factor Finder' tool. It would be easy to devise short activities for pupils using this resource. As pupils' access to the internet at home increases, it might be a worthwhile link to place on the school website, encouraging pupils to explore it in their own time.

Some teachers actively encourage pupils to see the mathematics that they are learning in school as part of a much, much bigger picture! Mathematics has been a human activity for thousands of years and the subject has a rich history of people, notation, images, theorems, laws, formulae and so on. If only we could produce a huge concept map of mathematics which could begin to link all these ideas!

WEBSITE

The Millennium Mathematics Project team is currently working on a multilingual project, funded by the European Union, to develop just this sort of map. The site is called 'Connecting Mathematics', http://thesaurus.maths.org/mmkb/view.html?resource=index.

Although it is best if you explore this site for yourself, here is an example.

Try typing the word 'rhombus' into the search facility. A mathematical definition of the word is given along with associated words, such as 'parallelogram' and 'square', which can be expanded if desired. An image of a rhombus can be selected. You can also select a 'graph' of the word, which is essentially a concept map, as in the example below.

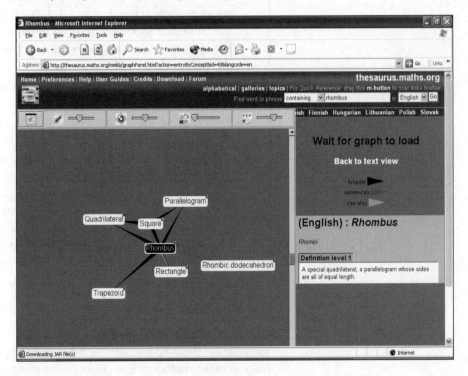

Selecting any of the words in the graph either expands the concept map further or, if you double click, makes the new word the central theme of its own concept map.

This is a phenomenal resource, guaranteed to open anyone's eyes to the amazing body of knowledge encompassed by the word 'mathematics'!

The project team give some suggestions for teachers and pupils, which link well with some of the approaches for the use of concept mapping software in the classroom discussed earlier in this chapter.

The thesaurus is a useful tool in planning, not only because it can help you check definitions of terms you are likely to use in a lesson, but also gives links to related topics and terms.

In your teaching you might try:

➡ preparing a graph of a mathematical concept in advance of a lesson, presenting it on a whiteboard or screen, and encouraging discussion about fitting a new or current topic into the pupils' current mathematical lexicon or world;

➡ creating graphs, and using definitions and ideas to include in worksheets;

➡ searching for a term, presenting the graph, then editing it. You could encourage pupils to discuss missing links or concepts, and delete concepts or links that are not of immediate relevance. The pupils can then view, even if briefly, the mathematics they are currently studying as part of a much larger whole.

Here are some ideas for pupils working in small groups, or independently:

➡ They could use the thesaurus to check the meaning of mathematical terms used in their books or by the teacher. The thesaurus will allow pupils to look at terms or words linked to their original enquiry. If they find that another term used in the definition they have found is not clear, they can use the references link to find an explanation of this term.

➡ Pupils might try to find out how a term links to other concepts they have met. The thesaurus tries to map these connections for them.

➡ They can use the thesaurus to produce a map of their own of an area of mathematics they are currently studying.

➡ Pupils could use the thesaurus to help them find material on other mathematics sites. Most sites have a search function, but it can sometimes be tricky to know what to type in. How do you find something when you can't remember what it is called? The answer is to use the thesaurus first! Start from something you know and use the relations to navigate to the term you want. Pupils can copy and paste the term into a search engine, or use the shortcuts to search engines provided.

In this chapter I have tried to take a wide view of 'making connections', focusing on the way we approach our pedagogy, and looking at ICT tools that help us make mathematical connections more explicit. We realize that we are all still learners of mathematics; this is never more obvious than when we come to teach the subject!

Chapter 6

Exploring and researching mathematics with ICT

In this section you will consider how:

■ ICT tools enable teachers and pupils to engage in mathematical explorations that are more open;

■ ICT tools encourage creativity in teachers and pupils;

■ ICT can be used to enable pupils to research mathematicians and areas of mathematics;

■ to source appropriate starting points for pupils.

If there is one question that gets all mathematics teachers and learners talking, it is the challenging question:

> do you think that mathematics is invented or discovered?

If mathematics is *discovered* by humans:

➡ who put it there?

➡ what is, and what is not, mathematics?

➡ have we, by now, discovered all mathematics? ...or a half, ...or a tenth?

If mathematics is *invented* by humans:

➡ how is it that people from different cultures have developed consistent ideas in mathematics?

➡ what is the explanation of the fact that the irrational numbers π and e are connected by the equation $e^{\pi\sqrt{-1}}=-1$?

➡ why do we keep coming across sequences of Fibonacci numbers in nature?

Irrespective of your personal view, exploring and researching mathematics is an exciting prospect. Your goal may be to discover new mathematics or to reconsider mathematics that has already been invented and perhaps invent some new mathematics! Either way it is a win-win situation.

Here is another question contributed by Afzal Ahmed that might provoke some more thinking on this theme:

'If people or beings from another planet were to observe our world, would they understand it in the same way that we do, or would it depend on the way their receptors (eyes, ears or whatever) are constructed?

Or, if we stay within our world, do ants see a quadrilateral as a quadrilateral?'

Your responses to these challenging questions may be linked to your religious and cultural beliefs; for you it may not be a question of invented or discovered at all!

Try asking colleagues and pupils what they think!

A common feature of most of the examples already described in this book is that teachers and pupils are using the ICT tools in an exploratory way. This is not by coincidence; the ICT tools have been deliberately chosen to facilitate such exploration. This is for a good reason; I believe that we learn mathematics by continually playing around with symbols, images and objects until we make our own mathematical sense of them. If we are playing, we need toys to play with!

When we are exploring mathematics using ICT the software becomes a test bed for mathematical ideas. Teachers and pupils can ask 'If we change … what will happen?' in a range of different mathematical contexts and settings. The element of prediction provides opportunities for teachers to assess pupils' prior knowledge and understanding, and for pupils to make their own inferences and construct their own mathematical meanings. The importance of a classroom culture in which both teachers and pupils ask their own mathematical questions has been discussed in Chapter 2.

A case study

Nicola Pruden, a Key Stage 3 mathematics consultant in Essex LEA, undertook some classroom research into how dynamic geometry software can support pupils to develop geometrical reasoning. While she was interviewing some Year 8 pupils, who were working with dynamic geometry software, one girl said,

'A long time ago I remember we ripped the corners of a paper triangle and had to fit it together on a straight line but mine was too big so I folded one bit over to make it fit and I didn't believe that it always worked because mine didn't.'

This is a good example of how one counter-example is sufficient to invalidate a mathematical proof and, as far as this pupil was concerned, she had found it!

Nicola observed the pupil working at the computer and noticed how she systematically proceeded to vary each of the interior angles of the triangle between zero and 180 degrees in an attempt to find the one counter-example that she had encountered previously. Only when her search did not yield a result did she focus on a different approach, based on her understanding of the angle properties of straight and parallel lines.

cont.

When Nicola asked the pupil how she thought the software was helping her develop a proof that the angle sum of a triangle is 180 degrees, the pupil said,

'Being able to play with the diagram, the proof was really easy to see because if you understand about alternate and corresponding angles they will always be relatively in the same places.'

In the United States, Nathalie Sinclair of Michigan State University has been developing a collection of activities, 'Sketchpad for young learners', which offers exploratory environments for pupils in the US equivalent of Key Stage 2. All these resources can be freely downloaded from

WEBSITE

http://www.keypress.com/sketchpad/general_resources/classroom_activities/young_learners/index.php

One of these activities that I particularly like is 'RooBooGoo', which gives pupils a draw tool and simultaneously shows an image that has been produced as a result of an unknown transformation.

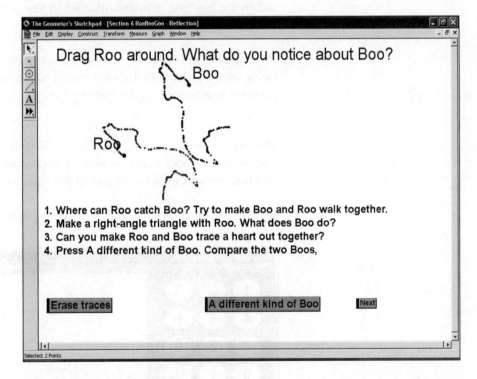

In the accompanying teachers' notes Nathalie writes 'The idea of this activity is to introduce students to some of the basic transformational behaviours. The emphasis is less on properties than on relationships between objects that have been reflected, rotated, or translated.'

In this activity pupils get the 'feel' of moving points that have been constrained in some way.

A major advantage of letting pupils explore mathematics in pairs is that it is very likely that, if the task is engaging, they will discuss the mathematics that they encounter.

If you can, download this activity and give it to a pair of pupils to explore, while you *just* observe them and listen. Don't intervene in any way…

What do they do?
Do they begin to make conjectures?
How do they begin to explain what is happening?
Do they use conventional mathematical language, or invent their own?
Do they set their own 'What if…' scenarios?
Do they ask their own questions?

Now, having observed the pupils working, what questions would you ask them in order to:

- confirm what you think they already understand;
- challenge them to think more deeply about their observations;
- develop their mathematical language;
- begin to assess what learning has taken place during the activity;
- move them on in their thinking?

The approach taken in this particular task could be adapted to complement the activities that were suggested in Chapter 1 when considering progression within the teaching of transformations in Key Stages 2, 3 and 4.

In Chapter 4, the dynamic number line on page 73 is a good example of ICT being used in a way that allows pupils to explore the relationships between numbers represented on the number line.

The tool 'Working with equations', which is part of the Multimedia Mathematics School resource, offers a responsive environment in which pupils and teachers can explore the role of the equals sign.

Initially, an equation such as the one below can be set up:

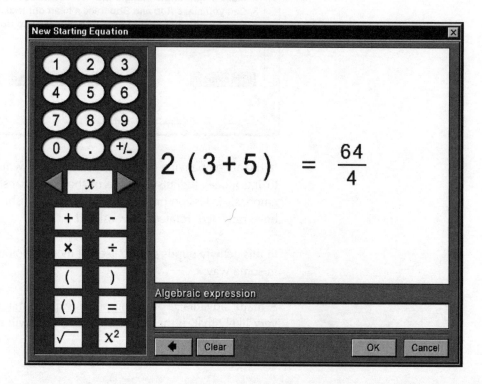

In most classrooms, displaying an equation such as this may immediately provoke some questions from pupils:

How can it be an equation – there are no letters?

What does $\dfrac{64}{4}$ mean?

What do the brackets mean?

In the environment created by this particular tool the whole notion of solving equations relies on finding equivalent expressions. Pupils could be encouraged to gain experience of finding equivalent expressions before they begin to develop a correct sequence of equivalent expressions that might lead them to solve an equation with one unknown.

Another feature of this particular ICT tool is that it allows the pupil or teacher to set a 'target' expression, which pupils then aim to reach. That is, pupils could be set a starting equation,

$$x = \dfrac{2y + 7}{5}$$

and a target equation.

$$7 = 5x - 2y$$

Pupils would then use the software to explore ways of transforming the equation. The feedback from the software is given at the point at which the pupil generates each subsequent line. The software is also suggesting how pupils might set out their work on paper.

An obvious approach, that teachers can adopt, is to challenge pupils to create their own equations that they will try to solve. By doing this they create a fund of their own experiences of dealing with particular types of equations, and begin to develop efficient routes for finding solutions. In this environment the pupils can play with equations informally, enabling them to arrive at their own approaches. This approach appears to have a lot in common with methods that are currently being advocated for teaching calculation strategies in Key Stages 1 and 2. Pupils are encouraged to develop their own strategies first, before exploring the most efficient way to arrive at a desired outcome.

CD-ROM

A 30-day trial version of Multimedia Mathematics School is on the accompanying CD-ROM – you can explore some equations for yourself!

Adopt, adapt, create...

Many teachers are happy to work with somebody else's pre-prepared script, in what is becoming labelled as the 'adopt' mode. But there are some who would like to be able to move towards the 'adapt' mode, in which they take into consideration pupils' existing knowledge and understanding by amending and deleting information, images or questions, or adapting activities in some other way.

This is often where frustrations set in; you would like to be able to adapt, but either the software has locked you out or you need to develop your own ICT skills to be able to adapt.

In Chapter 1, we focused on Paul Goldenberg's six principles and the 'fluent tool user'. If you have decided that a particular ICT tool is going to be valuable across the curriculum, age range and Key Stages, then you will need to invest some of your own time in getting to grips with it. Very often choosing to adapt a resource is an effective initial strategy.

The third stage, 'create', is reached when users move on to developing their own approaches and resources. This brings us back to creativity, which was first mentioned at the end of Chapter 1.

Creativity and ICT

How many times in a teaching day do you use the word 'create' in your mathematics classroom?

Some examples:

- ➡ create as many division calculations as you can with the answer 15;
- ➡ create an equation that gives a straight line through the point (3 , 5);
- ➡ create 4 triangles with an area of 12 cm^2;
- ➡ create a set of 5 numbers with a mean of 14.5 and a range of 4.

What are some of the advantages of asking pupils to create their own mathematics from given starting points?

- – Pupils have a degree of ownership within the task.
- – It allows pupils to generate their own ideas, and very often teachers are surprised by this and by how much their pupils already know.
- – Pupils can be asked to extend the task by changing one of the variables in some way.
- – Pupils begin to see mathematics as a subject to explore rather than as a set of right answers.
- – Pupils gain the confidence to experiment with mathematical ideas and concepts.
- – Pupils are encouraged to discuss and compare their responses with those of others.

And what about any disadvantages?

Marking!!

Undoubtedly, however delighted you are that pupils have fully engaged in a task and worked enthusiastically for the full length of a lesson, they can generate a huge volume of work. And, of course, you want to reward the pupils appropriately by giving your full attention to their responses. This is a dilemma that requires careful consideration.

ICT can offer quick and efficient feedback to pupils, which may mean that they do not record much of their work on paper. However, the use of the ICT enables them to move forward in their own understanding. This may mean that while pupils are engaged in ICT-based tasks your role in the classroom changes. Because you are not going to have paper-based outcomes at the end, do you spend your time in the classroom listening to pupils, and making appropriate interventions by carefully prompting and questioning?

Consider the following example.

The challenge is to create an equation that gives a straight line through the point (3, 5).

To begin with you need to choose an appropriate ICT tool that will allow pupils to plot the point on the graph and then input an equation and see its graph drawn.

All graphical calculators and graph plotting software have these facilities. But, because the way in which you input the co-ordinates and equations varies between tools, you will probably want to review the software carefully to decide what is most appropriate for your pupils.

Suppose a pupil has plotted the point and now wants to try to draw some straight lines. Depending on the age and ability of the pupil, you might consider asking questions such as:

What can you tell me about the values of the x and y co-ordinates at this point?

Do you think that this point would be on the line $y = x$?
Why? / Why not?

Careful questioning and prompting is needed to prevent pupils going into 'guess' mode, in which they select equations randomly, rather than using feedback from the previous attempt in a 'trial and improvement' approach. Some teachers suggest to pupils that they make notes or jottings alongside their ICT based work to encourage them to reflect more deeply before moving on.

In Chapter 3 we looked at how bringing visual images from the outside world into the classroom can provide motivating starting points for rich mathematical explorations. Adrian Oldknow has many excellent examples of this on his website http://www.adrianoldknow.org.uk. It gives pupils opportunities to use mathematics with which they are already familiar to try to model situations, and provides teachers and pupils with contexts for developing more sophisticated models.

WEBSITE

One of the most exciting innovations is the technology that enables you to paste pictures into dynamic geometry software, and do mathematics on top of the picture. This technique was used in Chapter 3 to explore the various animal tilings.

The recent QCA report *Developing Reasoning Through Algebra and Geometry* has featured these approaches within the context of algebra.

Using an image that is featured in Chapter 7, of a slide in a children's playground, the following approach could be taken.

How could we find the gradient of the slide?

Pasting the picture into The Geometer's Sketchpad allows us to take this question further in a variety of ways.

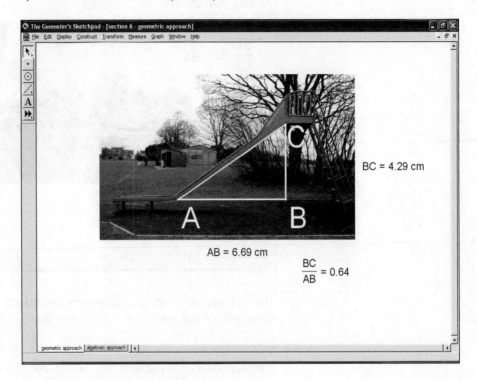

Before taking measurements and making calculations there are a few questions that need to be asked about this diagram!

– Do we really think that the length BC is 4.29 cm?
– What do we estimate that it should be?
– And what does that make the length AB in real-life?
– And does that mean that the gradient is bigger too?

I would ask pupils questions such as:

➡ How high do you estimate each step would be?
➡ Roughly how high does that make the slide?

Some fundamental ideas in the realm of scale drawings, similar triangles, enlargements and trigonometric ratios can be discussed using this one picture!

Alternatively, if you wanted to link this geometric approach with concepts that pupils might meet in algebra, you could consider overlaying some axes and scaling them to fit the real life situation.

Initially pupils may need to discuss where to place the origin and what would be a suitable scale for the axes. If the units are taken to be in metres, how does the scale need to be adjusted to fit the real-life situation? Adjusting the unit point on the *x*-axis then allows us to 'convert' the axes automatically.

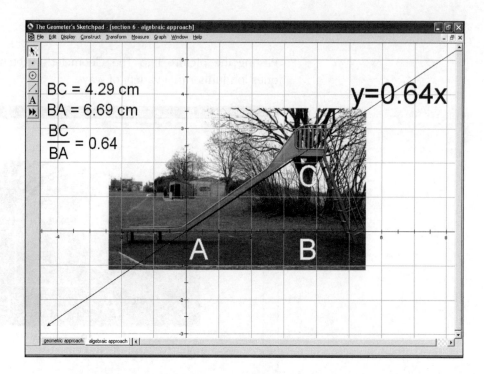

The calculation from the geometric approach can then support pupils to deduce the equation of the straight line through the origin.

This is another example of how the ICT is supporting pupils in making connections within mathematics and to the real world outside the classroom. There is nothing to stop this image being used to model the same situation in an AS mechanics or physics lesson, for example to estimate a sensible length for the run-off in order to prevent injuries in the playground!

▌ Researching mathematics

It would seem a very sad state of affairs if, within the 11 years of compulsory mathematics education, we cannot find time for pupils to carry out their own independent research into the life of a mathematician, a piece of mathematical history or a mathematical topic.

At Admiral Lord Nelson School in Portsmouth, all of the mathematics groups are named after famous mathematicians; Hypatia, Fermat, Pascal and so on. Early in Year 7, when pupils are making the transition into secondary school, they are set the task of researching their group's named mathematician's life and finding out about the area of mathematics to which he or she contributed.

Recently this kind of research has become fascinating as many websites devoted to the history of mathematics and mathematicians have been created. As a result pupils can uncover much more than basic biographical information. Also, the multicultural nature of the world wide web enables pupils to read accounts and interpretations from a non-Western world perspective. For example, pupils researching Pythagoras can learn that the Babylonian clay tablet, known as Plimpton 322, found in the Iraq desert, is evidence of an understanding of Pythagoras' theorem from about 1800 BC, which is more than a thousand years before Pythagoras! They could find out about Hypatia's untimely death, Fermat's famous last theorem, Pascal's love of gambling, and begin to place people at the heart of the development of mathematics. Mathematics is, after all, a human activity!

Just try putting 'history of mathematics' into a search engine to see where it takes you.

Internet health warning

The internet can be a very distracting media in which to work – how often do you find yourself 'hyperlinks' away from where you started just because something interested you, caught your eye or popped up! We could argue that this self-motivated learning is more valuable than the mere task in hand. However a certain amount of self-discipline is required!

Because a vast amount of information is available on the internet, we will all become more discerning about how we select appropriate or relevant information. Consequently pupils also need guidance about information retrieval. Tasks need to be precisely communicated to pupils, and designed so that there is a clear connection with the curriculum and pupils' existing knowledge.

This type of research activity could be supported by an adaptation of an 'advanced organizer'. The American psychologist, David Ausubel, first proposed the idea of an advanced organizer as a way of encouraging pupils to organize their thoughts about what they already know, before being introduced to the details of new concepts. When pupils are set clear tasks that will 'build' new knowledge an advanced organizer provides a place for them to record given starting points, references and deadlines. It is important to mention that David Ausubel believed in a didactic teaching style; he devised the concept of an advanced organizer to help pupils learn new factual information and not with the intention of enhancing pupils' understanding of any connecting concepts. He believed that, if pupils are given the big picture at the beginning of a task, they would be better able to take on the new knowledge.

Some general features of an advanced organizer are that it:

➡ has a clear introduction which places the context of the task firmly within the pupils' own experiences;

➡ gives an outline of the task itself;

➡ makes clear to pupils the links to their existing knowledge and understanding or work already covered;

➡ states the aims of the task;

➡ states clear objectives for the task;

➡ helps pupils organize their work – set deadlines, decide how the task should be presented and so on;

➡ includes resources that might be needed;

➡ provides space for pupils to write their own questions and allows them to develop the task further;

➡ explains any key words and has space for pupils to add and explain new mathematical vocabulary;

➡ provides space for pupils' own comments and reflections on their progress through the task.

If the research itself is to be carried out using ICT, the most helpful form for an advanced organizer is as an electronic document that pupils can personalize as they are working, and from which they can hyperlink.

All of this is best illustrated with a real example, which is what follows.

Sourcing appropriate starting points

WEBSITE

I have long been a fan of the cartoons by Tim Hunkin which were originally in *The Observer* colour magazine. All of these cartoons are now on a website http://www.rudimentsofwisdom.com.

For example, his cartoon 'Numbers' is a great starting point for a host of different research topics.

Irrespective of the Key Stage in which you are teaching, have you ever asked pupils why they think we use our particular number system? Or have you ever asked yourself that question?

Researching number systems from other cultures is a good example of an activity that will enable pupils of all ages and abilities to reflect upon some of the benefits of our system and how they make calculations easier. It also lets them see how mathematics has evolved throughout history, and the influence that ancient cultures and civilizations have had on our mathematical knowledge.

The Roman, Babylonian and Egyptian number systems are rich sources; pupils can explore similarities and differences with our own system.

Have a look at the following websites – these were all within the first five sites found using a UK-based internet search. There are hundreds of others which offer information and features in a variety of presentation styles. Some feature very advanced mathematics, others are most appropriate for primary age pupils.

WEBSITE   http://www.eyelid.co.uk/numbers.htm

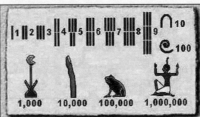

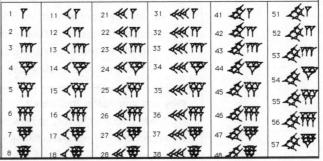

The Egyptians had a decimal system using seven different symbols.
1 is shown by a single stroke.
10 is shown by a drawing of a hobble for cattle.
100 is represented by a coil of rope.
1,000 is a drawing of a lotus plant.
10,000 is represented by a finger.

WEBSITE   http://www-gap.dcs.st-and.ac.uk
/~history/HistTopics/Babylonian_numerals.html

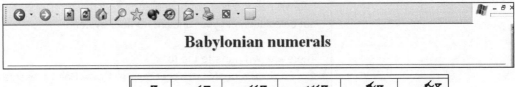

Babylonian numerals

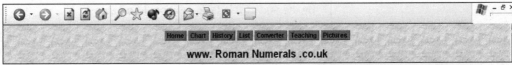

WEBSITE   http://www.romannumerals.co.uk/

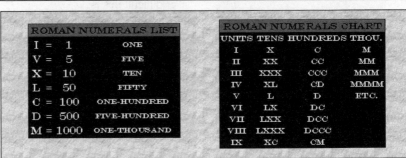

These three sites all give the 'code' from which pupils can begin to convert numbers. Some sites have conversion tools to provide feedback.

The next step is to use an advanced organizer to structure a task that can draw from such resources, and meet the criteria outlined previously.

EXPLORING NUMBERS FROM OTHER CULTURES AND CIVILIZATIONS

Introduction
Have you ever wondered why the number system that we use is based on tens, hundreds and thousands?
It is not the only number system that humans have developed.
You will probably have come across Roman numerals. Some watches and clocks still display the numbers in this way. You may also have seen Egyptian hieroglyphics.

The task
In this task you are going use the internet to research a number system from another culture or civilization.

You will need to find out how numbers were written and how you might carry out calculations in your ancient number system.

There are two parts to the task.
1. Produce a poster, concept map or PowerPoint presentation to explain to others how your ancient number system works, how to convert from decimal numbers and how to carry out some simple calculations.
2. Lead a discussion with a small group of other pupils in which you discuss the strengths and weaknesses of your ancient number system.

Why are you doing this task?
By looking back in history to how other cultures and civilizations used to work with numbers you will get a sense of how our decimal number system has developed. You will also think about the similarities and differences between some of the ways in which we carry out calculations.

Objectives
By the end of this task you should be able to:

- convert numbers from one number system to another;
- make calculations in your number system and comment on how easy or difficult this is;
- teach another group of pupils about your number system;
- lead a discussion about a mathematical topic.

Organizing your work
Make a note of deadlines, how you are going to present your work and so on.

Resources that you might use

Websites:

http://www.eyelid.co.uk/numbers.htm

http://www-gap.dcs.st-and.ac.uk/~history/HistTopics/
      Babylonian_numerals.html

http://www.romannumerals.co.uk/

Add any of your own here:

Key mathematical words

Word	Explanation

Write any of your own questions that you want to try to answer here

Evaluate your work here

How successful do you think you were in this task?

What difficulties did you encounter?

What new mathematics did you learn?

What would you do differently next time?

CD–ROM

Appendix 5 gives a template for an advanced organizer which is also available on the CD-ROM *Chapter 6 – advanced organizer.doc* as a Word document that can be adapted for your pupils and chosen theme.

As schools are developing the use of their websites, it is worth thinking about some creative ways of encouraging pupils to explore the history of mathematics through semi-structured approaches. These could be offered as recreational activities hosted on the school website for pupils to access in their own time at home. Is this the future of real 'homework'?

Chapter 7

Communicating mathematics with ICT

This section will look at issues and activities relating to communicating mathematics in a range of different contexts and using ICT tools, including:

■ producing written mathematics for paper-based resources;

■ pupils writing mathematically within an interactive environment;

■ using ICT to support pupils to communicate about mathematics;

■ sharing mathematical ideas in virtual learning environments.

Communication, the 'C' in ICT, is at the heart of teaching and learning mathematics. Teachers think carefully about how they will introduce and develop mathematical activities to support pupils to learn. Initially, in the Nursery or Reception classrooms, most of this communication is oral, aural and visual, supported by practical resources, while in Key Stage 4 classrooms, a fly on the wall would observe an abundance of aural and written communication. Historically, we have moved from clay tablets to paper and pencil as our main means of recording mathematics. ICT brings with it basic tools such as word processors which support us to create resources of a high presentation standard. There is also a range of more sophisticated tools, which we will look at later in the chapter, that have the feel of a word processor with a mathematical brain!

Using a basic word processor, such as MS Word, to communicate mathematics

Many of you will have presented tasks for pupils using MS Word. Letters, numerals and simple symbols such as + and – are all on the keyboard. You can find more mathematical symbols such as ÷ and √ from the Insert ⇨ Symbol palette, and, using superscript fonts, you can write indices such as cubed (^{3}) notation.

However, if we are to be critical, what can be frustrating? Possibly, trying to:

➡ add diagrams that stay where you want them on the page;

➡ include tables and graphs that look the way you want and stay where you want them on the page;

➡ write fractions so that they appear vertical;

➡ write more sophisticated algebraic expressions and formulae.

▍ Getting diagrams into Word documents

The Insert menu on the MS Word toolbar allows you to import a picture, either from another file to which you have access, or from the Clipart library. Most of us will do this fairly routinely to brighten up worksheets and set real-life contexts for tasks.

There are some things you can do, relating to inserting and formatting images, to stop images jumping around the page and losing their proportions when they are resized.

Take the following image, which was drawn accurately using The Geometer's Sketchpad software and copied and pasted into Word.

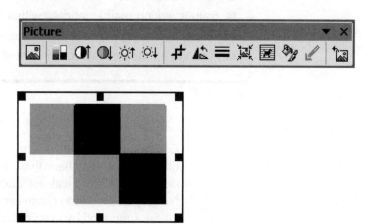

Having pasted an image into your document, when you select it, a Picture formatting toolbar appears. (This toolbar can float on the page, as shown here, or it can be dragged to the top toolbar, or it may already be there.)

The square markings on the edge of the image box allow you to resize the image. Selecting and dragging any of the corners will maintain the proportion of the image, while selecting and dragging the other marks will stretch the image either horizontally or vertically, depending on your selection. It is useful to have more control over this process, especially as it will support you to use images in a mathematical way, for example to produce a resource in which pupils can explore photographic enlargements.

The Key Stage 3 Strategy for Mathematics has produced a resource sheet called 'Cat faces' as part of the 'Interacting with mathematics in Year 9: Proportional reasoning materials'.

On this page, the images have been enlarged, some in direct proportion and some by horizontal and vertical stretching. The task for pupils is to discuss, hypothesize and check which of the images are mathematically similar and why.

A resource such as this is easily produced in MS Word.

If you are at a computer, open a new Word document and try it now! You may want to make a note of your enlargements as you go; these notes can be a set of answers.

Begin by copying and pasting an image onto the page… This is your 'original image'. (Refer to page 121 if you are not sure how to do this.)

You will probably want to be able to move the images around on the page while you are designing the worksheet. Usually, when you paste an image into MS Word it is placed in line with the text and you move it around using the return key or space bar. To do this, select the picture and click the format icon to open the 'Format Picture' window.

Select the 'Layout' tab.

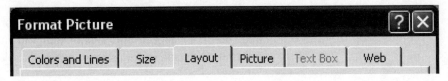

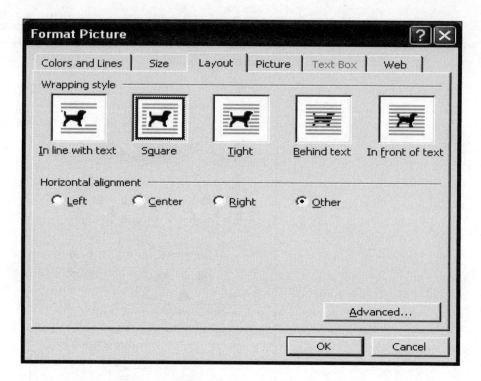

Select 'Square' or 'Tight', and you will then be able to drag the image anywhere on the page. Any text that you add to the page will be wrapped around the images.

Make a second copy of the image, select the image and open the 'Format Picture' window. This time, select the 'Size' tab.

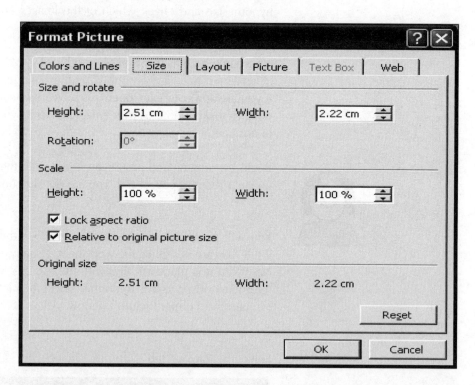

If you intend to make accurate enlargements of the original image, you need to keep the 'Lock aspect ratio' box ticked. You can then alter the enlargement by adjusting the scale factor as a percentage or selecting a specific height or width.

Fixing the scale at 40 per cent produced the following image.

If you want to stretch the image in some way, un-tick the 'Lock aspect ratio' box first. Choosing the width to be 6.4 cm produced the following stretch of the original image.

You can import images (with copyright permission) from many other sources, such as web pages.

There are several ways of doing this:

➡ If you position the cursor over the image a pop-up window appears with four icons that enable you to save, print, email or copy the image file as shown below.

➡ Position the cursor over the image and right-click the mouse to reveal the menu.

While it is easy to just copy and paste the images, it is also useful to begin to build a library of digital images for use in mathematics. This means you will need to save each one to a folder.

The slide images above were produced by Richard Phillips to support some teaching resources that have been produced by The Mathematical Association which exemplify how ICT can support differentiation in Key Stage 3 mathematics lessons. This particular lesson, 'Slopes', written by Joan Zorn, uses these images to develop pupils' understanding of slope and gradient. Many of the resources and teaching approaches could be easily adapted for pupils at Key Stages 2 and 4.

In Joan's lessons, she has pasted the slide images into dynamic geometry software which allows her to annotate accurately over the top of the image, to measure lengths and discuss slopes.

WEBSITE

These resources are freely available on the Becta website at: http://www.becta.org.uk

▌ Inserting tables and graphs

As there will be many occasions when you will want pupils to complete a table of information or work with a grid or graph, it is useful to be able to insert tables, grids and graphs into your document. Again MS Word can be frustrating when you are trying to position and format tables to look exactly as you want them to look.

All the following diagrams have been produced as 'Tables' in MS Word!

X	40	7
20		
9		

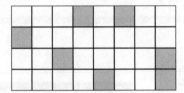

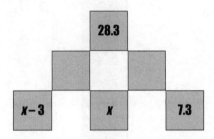

Here are some useful ICT tips:

➡ To draw a table, use the 'Insert' option from the 'Table' menu. In this way you can specify the number of columns and rows. If you deselect the 'AutoFit to Window' option, you can then specify the width of the columns in centimetres.

➡ To change the formatting for an individual cell, that is, to shade a single cell, select the cell by clicking near its left-hand edge. The cursor changes to an arrow, ↗, when you are in the right place.

➡ When you position the cursor above a table, a symbol appears in the top left corner. Click on this symbol and the whole table is automatically selected. A right click with the mouse gives the option to 'Distribute Rows Evenly' (or columns).

➡ Use border lines creatively. You can choose different line styles, colours and thickness for individual cells by selecting the cell and opening the 'Borders and Shading' window from the 'Format' menu on the toolbar.

➡ Discover the CTRL Y combination of keys. This repeats the last 'action' that you did. So, for example, if you are shading in the cells to make a fraction grid like the one above, shade the first cell using the 'Borders and Shading' window as above, but to shade subsequent cells, click in the individual cell that you want to shade and hold down the CTRL key while pressing the Y key.

▌ Using Equation Editor to write fractions and more sophisticated expressions and equations

Writing numerical and algebraic expressions and equations, and writing fractions vertically, require you to have 'Equation Editor' installed on your computer. Unfortunately, many computers do not have it installed because the Standard Installation for MS Office does not automatically install 'Equation Editor'.

When using 'Equation Editor', if you want, for example, to type the fraction 'two thirds', select 'Insert ⇨ Object' and choose 'Microsoft Equation' from the list. This will bring up a palette from which you can choose the format.

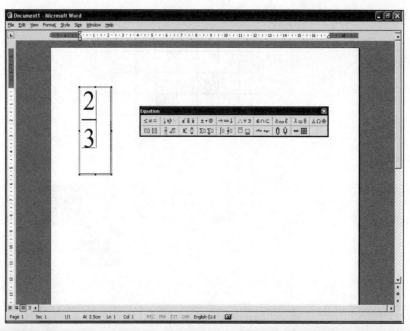

If 'Microsoft Equation' is not on this list, it is not installed on your computer. Type 'How do I install Equation Editor?' into MS Word 'Help', to get full instructions of what to do next. You may need to have a copy of the MS Office installation CD-ROM; 'Equation Editor' is one of the 'Office tools'. If you are working on a networked computer or do not have 'administrator' rights, you will probably need to enlist the help of an IT technician or the network manager.

There are other useful 'add-ins' to MS Office that allow more sophisticated mathematical images, equations and graphics to be produced.

WEBSITE

A popular one is 'FXDraw' by Efofex software. (A free 30-day trial version can be downloaded from http://www.efofex.com/)

The range of built-images, such as the number line tool and trigonometry diagram from the gallery, is extensive while still giving you the 'building blocks' to create your own images and save them for future use. Here are some examples:

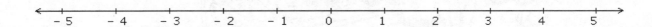

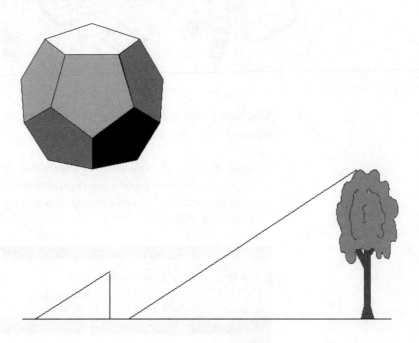

$$\frac{(2x^2 - 4x) + 2(x - 1.5)}{2x + 1} = 7$$

Both 'Equation Editor' and 'FXDraw' support teachers and pupils to write mathematical text and include it in word-processing software. However, a report written in MS Word will not be mathematically intelligent!

Typing '2 + 3 = 7' will not return any response from the computer!

It is now possible to use integrated mathematics software that is both a word processor and has a built-in mathematical brain.

One such piece of software is TI Interactive, produced by Texas Instruments. This has the appearance of a word-processing package, but also has a range of built-in mathematical tools to support teaching and learning number, algebra and data handling.

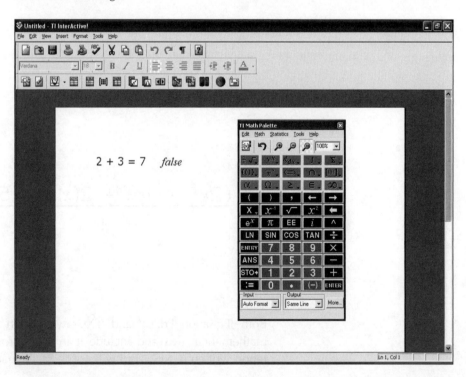

When using TI Interactive, if 2 + 3 = 7 is typed, the response 'false' is returned by the software.

TI Interactive provides an interesting environment in which pupils can begin to explore the role of the 'equals' sign. For example, given a starting equation, how many other correct statements can pupils generate?

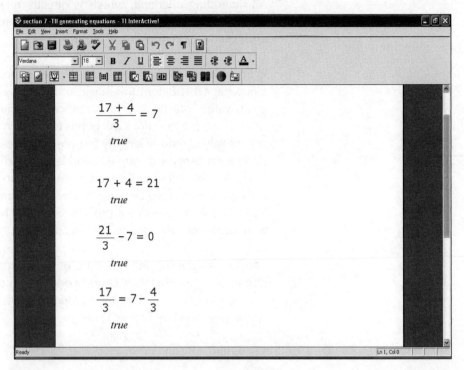

And which sequence of correct statements might lead to the solution of an equation?

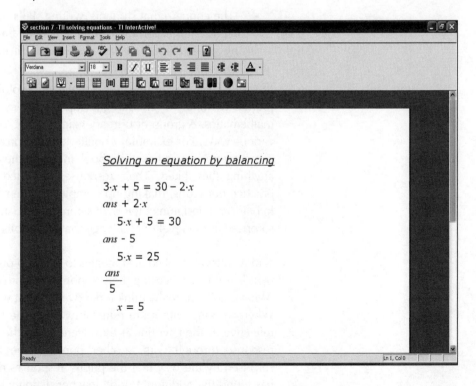

In both of these examples, the software is not 'giving the answer'; it is just responding to whatever is typed in using the calculator keypad. It is providing a test bed for pupils' emerging mathematical ideas.

▌ What is the value of mathematical communication in the school curriculum?

An interesting dilemma, which is directly related to how we communicate mathematics, arises as a result of our current mathematics curriculum and its associated national assessment framework.

In England and Wales, pupils usually begin their GCSE mathematics course at around age 14. Before this stage in a pupil's career there has often been no great value placed on communicating mathematically. Some Standard Assessment Test questions ask pupils to 'explain' an answer, but such questions are unlikely to require lengthy responses. Some schools will have been actively developing aspects of pupils' confidences and abilities to work on extended tasks. But, the skills pupils need when communicating sustained mathematical tasks are undervalued because they are not formally assessed as part of the Key Stage 1, 2 or 3 assessment process. So, what happens when pupils are faced with GCSE coursework and its expectations?

Candia Morgan of the Institute of Education in London studied pupils' responses to investigative coursework tasks, publishing the outcomes of this research in her book *Writing Mathematically: The discourse of investigation* (1998) published by Falmer Press.

She writes:

➡ 'In spite of increasing interest in the role and development of oral language in mathematics education, little attention has been paid to the role of writing.'

➡ 'The concern that students' writing skills may not be adequate to represent their problem-solving activity has not been accompanied by any attempt to enhance the writing itself.'

More recently, many teachers have been exploring the role of mathematics journal writing to encourage pupils to develop their writing skills in mathematics. A group of primary teachers in Croydon LEA approached this in various ways. For example, Annette Johnson provided each of her pupils with a centimetre-squared exercise book in which they had complete freedom to do anything they liked. Their responses included setting their own drill and practice questions, making up complex mathematical games that they played in pairs and designing their own symmetrical patterns. In all cases Annette was surprised at the level of challenge that the pupils placed on themselves!

Tammy Stewart looked at strategies to develop pupils' abilities to evaluate their own learning by focusing their responses towards answering questions such as 'What made you really think today?' and 'What was it that you found difficult?'. When working with more reluctant writers she found that they became more reflective in their writing if they were given the opportunity first to talk their response through with another pupil or the teacher. Again, Tammy was surprised by the way that the pupils' responses developed over time, and she has found the additional pupil-teacher dialogues invaluable as an assessment strategy.

Cara Ansell used mathematics journals to support her to find out more about how pupils learned from the visual, audio and kinesthetic teaching strategies that she was developing. She encouraged pupils to write about how the various models and images that she was providing helped them to learn.

A common aspect of these teachers' research projects was that pupils needed time to develop the way in which they used their journals. In all cases the teachers were disappointed by the pupils' early responses, but, with encouragement and support, the pupils gradually took ownership of their journals and began to value them as an important dialogue with their teacher.

In 2002, a change in the coursework requirements for GCSE led to the following feedback from QCA in 2003/4. 'The introduction of the handling data coursework task for GCSE caused considerable concern among schools. The most significant problems were with applying the assessment criteria.'

The headings for the three strands of the assessment criteria, which link to the data handling cycle, reveal that communication is at the heart of these processes.

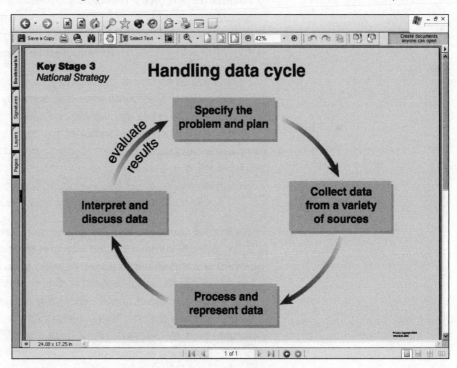

Have a close look at the following assessment criteria for the GCSE Handling data task; notice the important words or phrases relating to communication!

Specify the problem and plan

Candidates **describe** an overall plan largely designed to meet the aims and structure the project report so that results relating to some of the aims are brought out.

The project report is well structured so that the project can be seen as a whole.

They **state** their aims clearly in statistical terms and select and **develop** an appropriate plan to meet these aims **giving reasons** for their choice. The project report is well structured and the conclusions are related to the initial aims.

Collect, process and represent data

Their writing **explains and informs** their use of diagrams, which are usually related to their overall plan.

They use appropriate diagrams for representing data and **give a reason** for their choice of presentation, explaining features they have selected.

Candidates **use language** and statistical concepts effectively in **presenting** a convincing reasoned argument.

Interpret and discuss results

Candidates **comment** on patterns in the data and any exceptions. They **summarize** and **give a reasonably correct interpretation** of their graphs and calculations. They attempt to relate the summarized data to the initial problem, though some conclusions may be incorrect or irrelevant. They make some attempt to evaluate their strategy.

Candidates **comment** on patterns in the data and **suggest reasons for exceptions**. They **summarize** and **correctly interpret** their graphs and calculations, **relate** the summarized data to the initial problem and draw appropriate inferences. Candidates use summary statistics to make relevant comparisons and show an informal appreciation that results may not be statistically significant. Where relevant, they allow for the nature of the sampling method in making inferences about the population. They evaluate the effectiveness of the overall strategy and make a simple assessment of limitations.

Candidates **comment** on patterns and **give plausible reasons** for exceptions. They correctly **summarize and interpret** graphs and calculations. They **make correct and detailed inferences** from the data concerning the original problem using the vocabulary of probability. Candidates appreciate the significance of results they obtain. Where relevant, they allow for the nature and size of the sample and any possible bias in making inferences about the population. They **evaluate** the effectiveness of the overall strategy and recognize limitations of the work done, **making suggestions** for improvement. They **comment** constructively on the practical consequences of the work.

My discussions with teachers who had experienced difficulties applying the assessment criteria revealed that, in many cases, the mathematical content was evident in pupils' work. However, because they had not made written or oral comments that explicitly connected their hypotheses, calculations and graphs, teachers did not assess that they could confidently award a high mark. These difficulties in giving pupils credit for their understanding are not restricted to data handling. Teachers regularly comment that pupils find the GCSE 'Interpret and discuss' strand problematic at all levels of attainment.

So whose problem is it? Is it the teachers' problem, the pupils' problem, or does the problem lie in the assessment system itself?

In many other subjects, ICT is helping pupils, creatively and independently, to produce coursework of almost professional standards. In media studies, pupils use desktop publishing software to produce marketing publications. In design and technology, pupils use industry standard CAD-CAM software to design and make products, and in music, digital audio is manipulated with ease to produce recordings.

WEBSITE

> **ICT health warning**
> Be aware that websites exist, such as 'Course Info' (http://www.coursework.info/), which host hundreds of specimen coursework answers that pupils can download. For example, 'Data handling project for Mayfield School' is one of the pieces of coursework available to buy!

So what about mathematics? Let's return to the integrated mathematics software package TI Interactive, mentioned previously. Using TI Interactive it is possible to set up an interactive document in which pupils can add data, graphs and their own comments within a structured framework to produce an extended piece of work.

This is best illustrated by an example. Let's take a simple hypothesis that could be made in Key Stages 2, 3 or 4, 'Taller children have bigger feet', or an extension to this, 'Girls tend to have smaller feet than boys, in relation to their height'. How can we use ICT to support pupils in developing their responses to statements such as these?

WEBSITE

Many teachers are discovering the Census at School website (http://www.censusatschool.ntu.ac.uk), which has been created by the Royal Statistical Society Centre for Statistical Information, based at Nottingham Trent University. This website aims to:

➡ Provide real data for data-handling activities across the national curriculum, including increased awareness of the uses of surveys and censuses.

➡ Increase integration of ICT methodology within teaching and learning.

This interactive document, set up by a teacher, shows how one pupil used the resource.

Name ....

Handling data investigation: A structured approach

Through this task you will learn how to write a report of a statistical enquiry based on your own hypothesis or thoughts about a situation or some data. You will be able to include your own tables, calculations and graphs to support your writing.

You will aim to:

=> find out about making a hypothesis and deciding what data you will need to test if it is true

=> find some data sources and extract what you will need to begin your enquiry

=> use ICT to include tables, graphs and calculations

=> report on what you find out and decide if your original hypothesis was correct

=> include lots of thoughts and explanations as you work to show your understanding.

The diagram below shows the 'Handling data cycle' which will help you to review where you are in the task and what you need to do next.

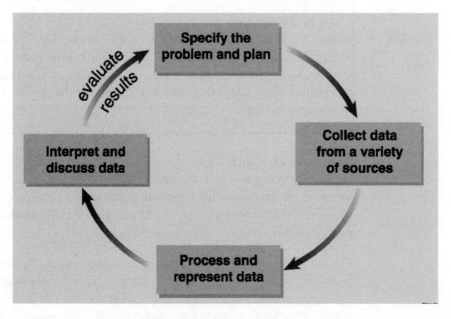

Write any important information here – deadlines, homework tasks, assessment and so on.

We have three lessons using the laptops and one homework to complete this task. It is due in on Monday 15 May. We can use websites to find suitable data.

We are being assessed on how we explain the task as well as the mathematics we use. We have to write down everything that you notice even if it does not seem ever so important.

Resources used in the task

Books: Maths Frameworking (Chapter on data handling)

Websites: Census at School website www.censusatschool.ntu.ac.uk

                    BBC Bitesize – revision of different types of averages and Box and Whisker graphs.

Other resources: tape measures, calculators

Vocabulary list

Make a note of any mathematical words and their meanings here:

mean average – when you share the values out equally between all of the data

range – the difference between the maximum value and the minimum value

mode – the most common value (or, if you group the data, the most common group)

median – the middle value when you put the data in order of size.

Developing your hypothesis

Record your initial hypothesis here:

I have decided to investigate two hypotheses

– that children who are taller have bigger feet

– that girls tend to have smaller feet in relation to their heights than boys.

How you are going to carry out the task

(include what data you will need to obtain, how you are going to organize it, what tables, graphs and calculations you will need to make and WHY...)

I am going to use a random set of data from the Census at School website – for about 500 school children who have taken part in the survey. I realize that I do not know if this is a representative sample as I don't know if the same amount of pupils in each age group have taken part in the survey.

I will separate the boys and girls data and calculate the range, mode, median and mean averages to see what it tells me.

I will need to compare these results against my hypotheses. I will do this by calculating some averages and comparing them. I will need to draw some graphs to show my data too.

When you are ready to input some data, go to Insert List to input a table and enter the values

You can change the names of the Lists by clicking on L1 and so on. (No spaces are allowed in list names and so on.)

You can copy and paste data into your table from another source.

I got the following data from the Census at School website – I copied it from an Excel file into a TI Interactive List so I could draw a range of statistical graphs more easily.

Gender	Height	Footlength
f	165	27
f	169	24.5
m	133	19
m	150	24.5
m	142	22
f	136	20
m	139	19
f	161	24
f	159	21
f	140	20
f	156	25
m	142	19.5
m	129	20
m	158	24
f	151	21
m	158	23
m	144	22
m	159	25
f	170	25
m	135	20
m	134	22
m	146	27.5
m	150	26.5
m	168	28
m	188	29
f	140	22
f	162	24
m	136	19
m	172	27
m	153	24

(NB. there are 500 data records in the real table)

What do you notice about the data in the table?

Does it give you enough information to decide if your hypothesis is correct?

What are you going to do next?

The data in my table is mixed up, male and female and it is not easy to make any conclusions from it.

I am going to separate the male and female data first.

Fgender	Fheight	Ffootlength	Mgender	Mheight	Mfootlength
f	165	27	m	133	19
f	169	24.5	m	150	24.5
f	136	20	m	142	22
f	161	24	m	139	19
f	159	21	m	142	19.5
f	140	20	m	129	20
f	156	25	m	158	24
f	151	21	m	158	23
f	170	25	m	144	22
f	140	22	m	159	25
f	162	24	m	135	20
f	175	23	m	134	22
f	136	22	m	146	27.5
f	137	20	m	150	26.5
f	152	22	m	168	28
f	161	23	m	188	29
f	158	18	m	136	19
f	172	24	m	172	27
f	136	22	m	153	24
f	140	22	m	140	26.5
f	153	24	m	170	22
f	155	26	m	155	23
f	172	27	m	138	23
f	162	24	m	176	27
f	154	23	m	142	22
f	157	23	m	151	25
f	161	25	m	152	27
f	164	24	m	163	25
f	124	20	m	147	22
f	123	18.5	m	179	29

I can see now that there are 283 females and 234 males in the survey. Ideally I would like to have the same number to be able to compare the two so I am going to ignore the last 49 females.

If it would be useful to make some calculations, which calculations will you make and why?

I will begin by calculating the range, mode, median and mean averages for each of the sets of data.

Female height data
One-Variable Statistics

$\overline{x} = 149.56$

$\Sigma x = 34997.$

$n = 234.$

$minX = 114.$

Median $= 152.$

$maxX = 179.$

Male height data
One-Variable Statistics

$\overline{x} = 150.226$

$\Sigma x = 35153.$

$n = 234.$

$minX = 119.$

Median $= 150.$

$maxX = 188.$

I can see straight away that the male mean average height is greater than the female mean average by just over half a centimetre. I have drawn a box and whisker plot of the data. The top graph is the females and the bottom graph is the males.

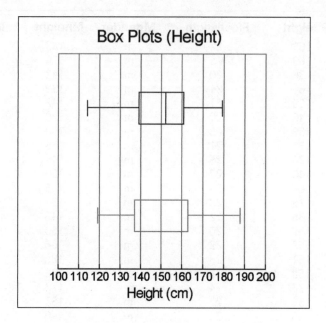

The graph clearly shows that the tallest boy (188 cm) is taller than the tallest girl (179 cm) and the shortest boy (119 cm) is also taller than the shortest girl (114 cm); however, the median average height is less for boys (150 cm) than it is for girls (152 cm).
Looking at the graph the box section for the girls is more compact than for the boys. This means that the girls' heights are closer together.

Look back at your hypothesis. Do the results of your calculations give you enough information to decide if your hypothesis is correct?

I am not sure, it depends which average you choose. I think that for a big set of data, like 500, it is best to look at the mean average as the calculation has shared out everyone's height equally. This means that my hypothesis is correct.

Would it be useful to draw some graphs?

I have already drawn some box and whisker plots to help me explain my first hypothesis. To investigate my second hypothesis, I will need to draw scatter graphs of the height against foot length for both females and males. I will try to find the gradient of the line of best fit. I am expecting the male graph to be less steep.

Which type(s) of graphs will you draw and why?

Select Insert and Graph to draw a graph in your document.
You will need to type the List names for the type of graph that you want exactly as they were in your tables.

By looking at both graphs, I can see that there is a strong positive correlation between the heights and the foot lengths for females and males. I am going to try to find the equation of the line of best fit. I know that it must go through (0,0) as height is directly proportional to foot length.

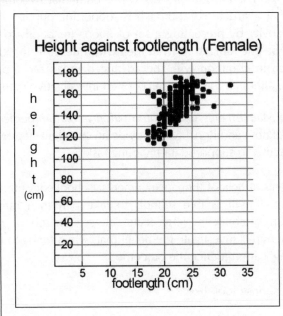

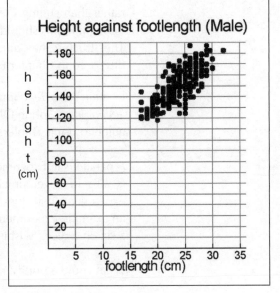

The equation of the line of best fit for the female graph is y = 6.5x

The equation of the line of best fit for the male graph is y = 6.3x

This means that the females' heights are roughly 6.5 times their foot lengths and my second hypothesis is correct.

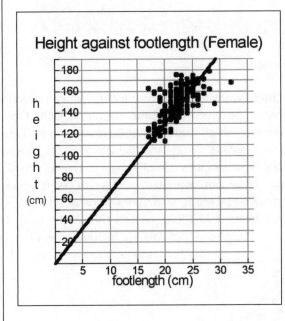

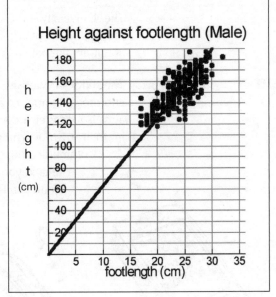

Evaluate your work

Have you got sufficient information to decide if your hypothesis is correct?

What would you like to do next, if you have the time?

I think that the data I collected did enable me to investigate my hypotheses; however, if I were going to do this task again, I might make sure that I had roughly the same numbers of boys and girls from each age group.

Obviously, the teacher designed this framework with a particular hypothesis in mind. It is important that a structured approach is supporting pupils' writing, rather than stifling creativity by presenting a set of hoops for pupils to jump through. Pupils should be encouraged to write their own questions to explore. The structure has much in common with the advanced organizer approach discussed in Chapter 6, although the task itself is of an investigative nature.

In their book *Teaching Mathematics Using Information and Communications Technology*, Adrian Oldknow and Ron Taylor provide an excellent step-by-step guide to using TI Interactive to support data-handling tasks.

WEBSITE

A parallel site Experiments at School (http://experimentsatschool.edev.ntu.ac.uk) is also in development. It will enable schools to take part in a range of experiments with the aim of collecting large-scale data sets for school use. In an example experiment pupils investigate the belief that, when trying to throw a six on a dice, 'wishing' can influence the outcome!

What could a pupils' response look like?

Pupils communicating about mathematics

In addition to the obvious verbal communications that happen inside and outside mathematics classrooms, ICT is already offering creative ways in which pupils can communicate mathematically through videoconferencing and on-line chat resources.

The University of Cambridge, has developed the NRICH and MOTIVATE projects, both of which offer ways for pupils to work on, and communicate about, mathematical problems.

WEBSITE

The NRICH website (http://www.nrich.maths.org.uk), which contains an enormous number of articles, problems and games, began in 1996. One of its aims is to 'foster a community where students can be involved and supported in their own learning and where effort and achievement is celebrated'. Communicating about mathematics is problematic if the only medium is a text keyboard. How can you share a diagram, sketch, graph or jottings? The NRICH website overcomes some of these difficulties within its 'askNRICH' area of the site by using HTML keyboard strokes to include specific symbols such as square root symbols and index notation.

It is a major consideration for all software developers to look at ways of devising handwriting recognition software that will recognize common mathematical notation and representations, especially with the arrival into classrooms of tablet PCs and interactive whiteboards. Alternatively, will we develop a new vocabulary of mathematical shorthand in much the same way that text messaging has produced a special shorthand 'vocabulary'?

WEBSITE

The MOTIVATE project (http://motivate.maths.org/) enables primary and secondary school pupils to take part in videoconferencing events on a wide range of mathematical themes. Some aims of the project are to:

➡ give pupils an experience of collaborative working on mathematical tasks and of presenting reports of their work to an audience;

➡ raise aspirations and to improve mathematical thinking and communication skills;

➡ help schools develop videoconferencing as part of the resources available to them;

➡ develop guidelines for technical and media presentation that will also serve other subjects across the curriculum;

➡ enable schools to develop videoconferencing links with other schools around the world.

Schools are supported to develop their ICT infrastructure so that pupils can participate in MOTIVATE conferences; in one model pupils experience two videoconferences, about a month apart, and work on a range of related activities in the interim period.

Warden Park School in Haywards Heath, the first school to achieve the status of being a languages, mathematics and computing specialist school, is keen to develop mathematical links with schools internationally. For example, pupils are intending to explore cultural similarities and differences with their peers at partner schools by using email to exchange lifestyle data. If you were to set up a similar innovative initiative, it would be a good idea first to investigate cultural differences between school curricula; the website of the International Review of Curriculum and Assessment Frameworks (http://www.inca.org.uk/) would be a useful starting point.

WEBSITE

▌The future: developing virtual learning environments for mathematics

An accepted definition of a virtual learning environment (VLE) is 'a set of components in which learners and teachers participate in online interactions of various kinds'.

Typically, a VLE might:

➡ provide access to a curriculum in a secure area;

➡ be able to track pupils' activity and achievement;

➡ provide online support in the form of resources, assessment and guidance;

➡ facilitate communications between pupils, teachers and support staff;

➡ be able to link to other administrative systems, both in-house and externally;

➡ be able to be customized.

Many school intranet sites are being designed with some of the features of a VLE. However, one of the most important features of a VLE is that all its users can develop it. Every member has the facility to upload files, design their own content and communicate with others on the site. VLE software gives the tools to the users.

At the time of writing, VLEs are being developed by partnerships of schools and LEAs. For example, 'E-Sy' is being developed by a group of South Yorkshire LEAs; Barnsley, Doncaster, Rotherham and Sheffield. They state that the project is about developing ICT skills across all its communities with a view to 'giving young people, job seekers and employed staff the competitive advantages they need to thrive in today's digital economy'.

As these projects progress it will be vital that any resources developed for mathematics encompass creative characteristics such as those exemplified in this book. It would be a great shame for users to log on to their VLE only to experience an online textbook that does not offer the range and diversity of these approaches.

Some questions to consider:

➡ What will your online 'curriculum for mathematics' look like?

➡ What software tools will you give pupils to use?

➡ In an online environment, how will pupils 'write' mathematics in a way that caters for a wide range and diversity of responses?

➡ How will you avoid the easy option of asking for 'Yes/No' or 'Correct/Incorrect' responses to online pupil tasks and, thus, compounding the view that doing mathematics is 'all about right answers'!

➡ Who will provide the important feedback to pupils?

It is appropriate to close this book with a look into the future.

What will the mathematics classrooms of 2025 look like?

Will pupils need to go to a particular place to learn mathematics?

How will the role of the teacher have changed?

In order to support pupils, will teachers need to know more or less mathematics?

Will pupils use any practical resources?

Will pupils have ceased to discuss mathematics, or will they have more opportunities to do so?

Add some questions of your own…

Appendix 1

Evaluating mathematical learning – Sort cards

Using mathematical terms	**Using mathematical notation**
Using mathematical conventions	**Obtaining results**
Performing basic operations	**Sensible use of a calculator**
Using simple practical skills in mathematics	**Communicating mathematics**
Using ICT in mathematical activities	**Understanding basic concepts**
Making relationships between concepts	**Selecting appropriate data**

Appendix 1

Evaluating mathematical learning – Sort cards

Using mathematics in context	**Interpreting results**
Using the ability to estimate	**Using the ability to approximate**
Employing trial and improvement methods	**Simplifying difficult tasks**
Looking for patterns	**Reasoning**
Making and testing hypotheses	**Proving and disproving**
Developing good work habits **Being imaginative, creative or flexible**	*Developing good work habits* **Being systematic**

Appendix 1

Evaluating mathematical learning – Sort cards

Developing good work habits **Being independent in thought and action**	*Developing good work habits* **Being co-operative**
Developing good work habits **Being persistent**	*Developing a positive attitude to mathematics* **Developing a fascination with the subject**
Developing a positive attitude to mathematics **Showing interest and motivation**	*Developing a positive attitude to mathematics* **Showing pleasure and enjoyment from mathematical activities**
Developing a positive attitude to mathematics **Showing an appreciation of the purpose, power and relevance of mathematics**	*Developing a positive attitude to mathematics* **Showing satisfaction derived from a sense of achievement**
Developing a positive attitude to mathematics **Showing confidence in an ability to do mathematics at an appropriate level**	

Appendix 2

Checklist for lesson planning

		Objective	✔	How the ICT will support pupils to achieve this in the lesson
Facts		Using mathematical terms		
		Using mathematical notation		
		Using mathematical conventions		
		Obtaining results		
Skills		Performing basic operations		
		Sensible use of a calculator		
		Using simple practical skills in mathematics		
		Communicating mathematics		
		Using ICT in mathematical activities		
Conceptual strategies		Understanding basic concepts		
		Making relationships between concepts		
		Selecting appropriate data		
		Using mathematics in context		
		Interpreting results		
		Using the ability to estimate		
		Using the ability to approximate		
		Employing trial and improvement methods		
		Simplifying difficult tasks		
		Looking for patterns		
		Reasoning		
		Making and testing hypotheses		
		Proving and disproving		
Personal qualities	Developing good work habits	Being imaginative, creative or flexible		
		Being systematic		
		Being independent in thought and action		
		Being co-operative		
		Being persistent		
	Developing a positive attitude to mathematics	Developing a fascination with the subject		
		Showing interest and motivation		
		Showing pleasure and enjoyment from mathematical activities		
		Showing an appreciation of the purpose, power and relevance of mathematics		
		Showing satisfaction derived from a sense of achievement		
		Showing confidence in an ability to do mathematics at an appropriate level		

Appendix 3(a)

Lesson plan: Real-life graphing

Before the lesson:	<ul><li>Organize the computer, motion sensor (CBR) and data projector for whole-class teaching.</li><li>Save the file **Resource M01** *Creating distance-time graphs.tii* to the shared area of the school network or onto the laptops for pupil access.</li><li>Organize laptops, CBRs and link cables for pupil use.</li><li>Print sufficient sets of **Resource P1** *Journeys to School* cards for one set between two pupils.</li><li>Print and make available the Texas TI Interactive **Helpsheet** Using the CBR for pupils.</li><li>Decide whether pupils will be required to print their work and, if so, make a printer available in the classroom.</li></ul>

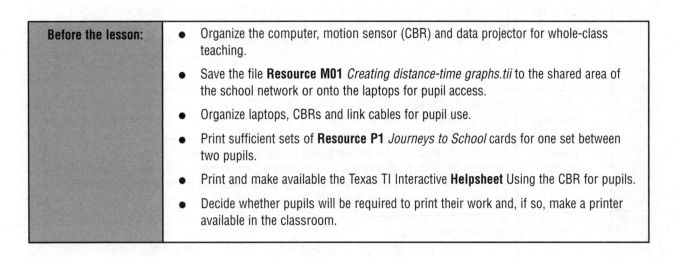

	Objectives:	Construct linear functions arising from real-life problems and plot their corresponding graphs; discuss and interpret graphs arising from real situations.
Oral and Mental starter	Vocabulary:	Distance-time graph, gradient, intercept, slope, steepness, horizontal, vertical
	Activity:	Display the Word file **Resource M01** *Graph Stories* – Give pupils a minute to discuss in pairs what 'real-life' situation the graph could represent (for example, water level against time, distance against time). Ask pupils to record and display their responses on individual whiteboards. Inform the pupils that it is a distance-time graph and ask them to discuss what the axes and scale could be. (The y-axis would need to represent the distance from a particular starting point and the x-axis would represent time. Distances could be metres, kilometres, miles, centimetres and so on. Times could be seconds, minutes, hours, days and so on.) Move to the second graph. Again, give pupils a minute to discuss their responses and display them on individual whiteboards. Discuss the fact that, if the y-axis represents distance and the x-axis represents time then the graph involves 'moving back in time'. However, if the y-axis represented the time, and the x-axis the distance, the graph could represent a distance-time graph. Explain that the usual convention is for the time to be represented by the x-axis and the distance by the y-axis. Any units of distance and time can be used; however, we choose the most sensible unit for the situation that is being described by the graph.
	Key questions:	<ul><li>What could the graph represent?</li><li>What could the axes represent?</li><li>What could the units of measurement be?</li><li>What 'story' could the graph be representing?</li></ul>
	Resources:	**Resource M01** *Graph Stories* – Whole-class computer display (interactive whiteboard or large screen) Individual whiteboards and pens

Appendix 3(a)

Lesson plan: Real-life graphing

<table>
<tr>
<td rowspan="8">Main lesson activity</td>
<td>Objectives:</td>
<td>Construct linear functions arising from real-life problems and plot their corresponding graphs; discuss and interpret graphs arising from real situations.
Identify the necessary information to solve a problem; represent problems and interpret solutions in algebraic, geometric or graphical form, using correct notation and appropriate diagrams.</td>
</tr>
<tr>
<td>Key vocabulary:</td>
<td>Distance-time graph, gradient, intercept, linear relationship, slope, steepness, horizontal, vertical</td>
</tr>
<tr>
<td>Teacher activity:</td>
<td>Open the TI Interactive file **Resource M1** *Creating distance-time graphs*.
Demonstrate how to set up the CBR and invite a pupil to come and walk a graph in real-time. Ask the pupil to move slowly and quickly, forwards and backwards and to stand still. Also, ask the pupil to wave, jump or turn around.
Discuss the resulting graph.
Ask questions such as:
    Which part of the graph corresponded to when … was moving away from the CBR?
    What about moving towards the CBR?   Or standing still?
    Could you tell when … was waving, jumping, turning? Why/why not?

Show pupils how to edit the appearance of the graph, adding a title and labelling the axes and the correct units (see the **Helpsheet** for details).
Show pupils how to add text to the document to describe in words the story of the graph.

Explain that, by the end of the lesson, pupils will be expected to each have produced their own graph and included some text which describes their movements.</td>
</tr>
<tr>
<td>Pupil activity:</td>
<td>Pupils will be working in small groups using a laptop and CBR.
Ask them to open the file **Resource M1** *Creating distance-time graphs* and use the CBR to create their own distance-time graphs.
Encourage pupils to explore different types of motion and the resulting graphs.
After each graph has been created, ask pupils to collaborate to write the 'story' for each graph, adding as much detail as they can. Pupils can use the distance-time data from the table to make more accurate statements such as, 'after 3.2 seconds, Sarah moved a further 2.1 metres from the CBR'.
When pupils have had the opportunity to create their own graph, give pupils an opportunity to discuss their responses to the questions in Resource M1. Hold a mini-plenary to expose common misconceptions such as, if the graph has a positive gradient, the person was moving uphill.
Give pupils some time to write a group response to the questions and save and their work.</td>
</tr>
<tr>
<td>Key questions:</td>
<td>
● What does the steepness or gradient of a distance-time graph represent?
● Is it possible to have a horizontal line on a distance-time graph?
● What about a vertical line?
● What information is it possible to obtain from a distance-time graph and what is not possible?
</td>
</tr>
<tr>
<td>Resources:</td>
<td>TI Calculator Based Ranger (CBR) and whole-class display
TI Interactive software
Laptops and CBRs for pupil use
Resource M1 *Creating distance-time graphs*
Helpsheet Texas TI Interactive – Using the CBR</td>
</tr>
</table>

Appendix 3(a)

Lesson plan: Real-life graphing

<table>
<tr>
<td rowspan="4" style="writing-mode: vertical-rl;">Plenary</td>
<td>Activity:</td>
<td>Give out a set of Resource P1 Journeys to School cards to each pair of pupils. Inform pupils that all three of the graphs have the same scale.
Ask pupils to discuss and decide which graph matches with which story.
Encourage pupils to annotate the graphs to support their decisions.
Ask pupils to underline any information in the text that is particularly useful and cross out any that is irrelevant.
After about five minutes, take feedback from pairs of pupils.
Discuss the strategies that pupils have used.
Discuss the misconception that some pupils may hold in that they associate Graph 2 with Chris as it is has the steepest lines and he lives at the top of a hill.
Answers
Laura = Graph 2
Key information: She stops four times. The train stops are fairly similar in length. The three periods when the train is moving are of a similar gradient, and the steepest of all of the graphs.
Irrelevant information: The time she leaves home. That she buys a magazine.
Chris = Graph 1
Key information: He stops twice on the way. He walks to school so distance will be near.
Irrelevant information: That he lives at the top of a hill. That he leaves home at twenty past seven. That he calls for two friends.
Mina = Graph 3
All of the information is key, including 'it might be quicker to walk' which prompts you to compare the average gradient with that of Graph 1.</td>
</tr>
<tr>
<td>Key questions:</td>
<td>●   What key information in the story is most useful when trying to match the journeys with their graphs?</td>
</tr>
<tr>
<td>Resources:</td>
<td>Resource P1 Journeys to School</td>
</tr>
<tr>
<td>Homework:</td>
<td>Give pupils a local map that will enable them to mark their route to school.
Ask them to write on the map the times at which they pass a few identified landmarks and draw a distance-time graph for their journey. You will need to discuss how pupils can estimate the distances travelled from the map using a piece of string and the scale.
NB. Ordnance Survey provides free maps of the school locality to all Year 7 pupils. Check with your geography department to see if the maps have been requested. If so, there may be a stock of local maps that you could use for this task. If not, the information is available from: http://www.freemaps4schools.co.uk/freemapsfor11yearolds/</td>
</tr>
<tr>
<td colspan="2">Assessment opportunities</td>
<td>During the mental/oral starter, use the key questions to assess pupils' prior knowledge and understanding of the interpretation of graphs.
During the pupils' main activity, use the key questions to establish what pupils have learned from the practical activity. If pupils have printed their work, assess whether they have understood how to relate the movements that they made with the resulting graphs.</td>
</tr>
</table>

Appendix 3(b)

Resource MO1 Graph Stories

What could this graph represent?

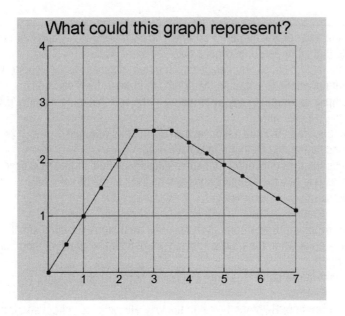

If this graph represented a distance-time graph, what *could* the axes mean?

What could the units of measurement be?

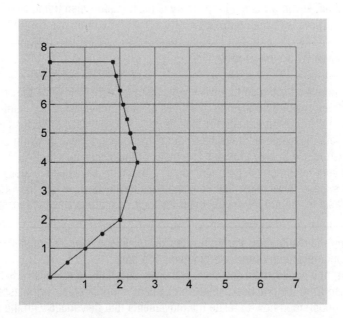

Could this graph be a distance-time graph?

Why / Why not?

Appendix 3(c)

Resource P1 Journeys to School

Laura leaves home at 07:00. She walks to the corner shop where she stops to buy a magazine. She arrives at the station, waits a short while for her train, which stops at two more stations before she arrives at the station nearest to school. She walks from the station to school.

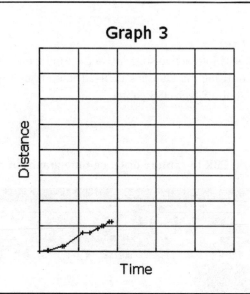

Chris lives at the top of a hill. He leaves home at twenty past seven in the morning to walk to school. He stops to call for two friends on the way.

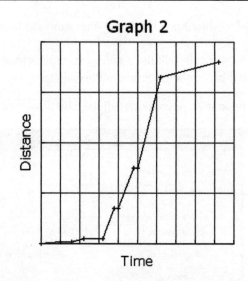

Mina travels to school by car. She lives in a busy town. It might be quicker to walk!

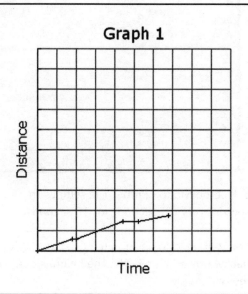

Appendix 3(d)

Texas TI Interactive Helpsheet – Using the CBR

Setting up

Open the TI Interactive software and connect the CBR to the computer using the USB link cable. The computer will 'find' and recognize the Silver USB cable.

Using the CBR to capture distance-time graphs in real time

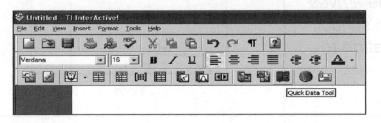

1. Select the Quick Data Tool icon from the toolbar.

2. The software will then attempt to connect with the CBR. If the device is not found, check that the USB lead is connected firmly and try again.

    The following window will then be displayed.

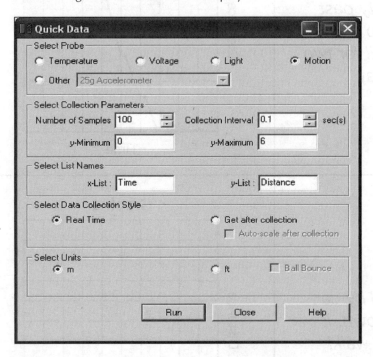

3. Select the Motion Probe. The settings shown will instruct the CBR to measure the distance from the sensor every 0.1 seconds for 10 seconds.

4. You will need to position the CBR so that it is pointing towards a space in which it is possible to move backwards and forwards. The beam is conical, so you will need to make sure that nothing is obstructing the wider view.

Exciting ICT in Maths

Appendix 3(d)

Texas TI Interactive Helpsheet – Using the CBR

5. When ready, Select Run. The CBR will make a clicking sound. As you move forwards and backwards in front of the CBR the points on the graph are plotted, showing your distance from the CBR and the elapsed time.

6. When the data has been collected the following screens will be displayed:

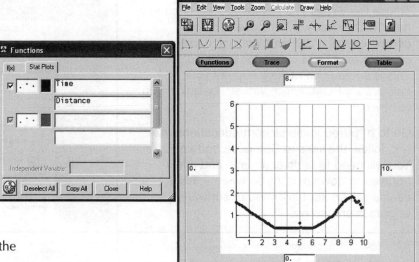

Select Format to change the appearance of the graph, label the axes and add a title.

Close these two windows to drop the graph into the document.

7. Close the Quick Data window and you will see the Data Editor window with the table of Times and Distances. As there is no data in Lists 1 to 6, select and delete them using Edit menu in the Data Editor.

8. Close the Data Editor window to drop the table and graph into the document.

Printing TI Interactive files

Before you print a TI Interactive file, you will need to save it as a text file. This is because TI Interactive files can be very large and take a long time to print.

Select Export from the File menu and choose RTF format. You will need to choose a filename and save this document to your own folder. You can now print to a selected printer.

Appendix 4

Helpsheet – Installing Virtual TI-83 onto your PC

1.

Go to http://www.ticalc.org/archives/files/fileinfo/261/26120.html
And download the Virtual TI software. Make a note of the folder in which the software is saved as you will need to save another file to this folder later on for the software to run properly.

2.

To use the Virtual TI software you will need to have TI Connect v. 1.3 installed on your computer and a USB or black link cable to connect your TI 83 calculator to the computer. This is so you can make a copy of your calculator's ROM, which Virtual TI requires you to have.

Go to http://education.ti.com/educationportal/ and follow the links to Computer software. You will need to register a name and email address to download the software – this is useful as you will be kept informed of software upgrades and so on.

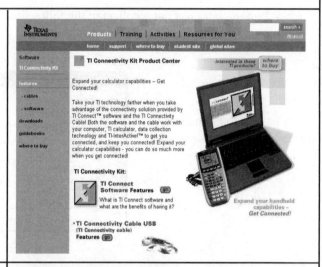

3.

Once the software is installed, open TI Connect, connect your calculator to the computer using a link cable and click the **Backup** icon.
You may be prompted to say which calculator and cable you are using.

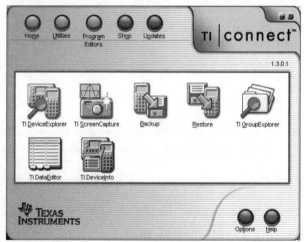

4.

Save the Calculator ROM file to the same folder as the Virtual TI .exe file.
When you open the Virtual TI file, you will open an on-screen version of your calculator, complete with any programs, applications or data that was on your own calculator.

Exciting ICT in Maths

Appendix 5

Template for an advanced organizer

Maybe a related image here...

All of this column provides 'white space' for pupils' annotations.

TITLE

INTRODUCTION
Introduce the task by giving pupils the big picture and then focus it towards the pupils' own lives and experiences

THE TASK
Provide a clear and concise description of what it is that you want the pupils to do.

WHY ARE YOU DOING THIS TASK?
This is where you would link the task to the curriculum and provide the purpose to the pupils.

OBJECTIVES
Give clear and specific objectives both in a mathematical sense but also referring to key skills, how you might except pupils to work and so on.

ORGANIZING YOUR WORK
This is where you would give pupils an idea of the timescale, which part will be done as homework, when and how the work is to be handed in for assessment and so on.

RESOURCES THAT YOU MIGHT USE
Include references to websites, CD-ROMS, books, videos and so on.

Invite students to add their own here:

KEY MATHEMATICAL WORDS

Word	Explanation

Appendix 5

Template for an advanced organizer

WRITE ANY OF YOUR OWN QUESTIONS THAT YOU WANT TO TRY TO ANSWER HERE

Actively encourage pupils to formulate their own questions to try to answer.

EVALUATE YOUR WORK HERE

How successful do you think you were in this task?

What difficulties did you encounter

What new mathematics did you learn?

What would you do differently next time?

The nature of any questions that you write here will focus pupils towards self-assessing how they have achieved in the task You will need to direct them back to the original objectives in order that they can develop effective self-assessment practices.

References

Mundhar Adhami (2001), *Support for Learning*, Vol. 1, issue 1, pp. 28–34

Afzal Ahmed and Honor Williams, (2002), *Numeracy Activities: Plenary, practical and problem solving*, Network Educational Press

Abraham Arcavi (2003), 'The role of visual representations in the learning of mathematics', *Educational Studies in Mathematics*, Vol. 52, p. 217

Mike Askew (1997), *Effective Teachers of Numeracy*, King's College London

Assessment Reform Group (1999), *Assessment for Learning: Beyond the black box*, University of Cambridge

Arthur Baroody and Bobbye Hoffman Bartels (2002), 'Assessing understanding in mathematics with concept mapping', *Mathematics in School*, Mathematical Association

Paul Black and Dylan Wiliam (1996), *Inside the Black Box*, King's College London

Oliver Caviglioli and Ian Harris (2001), *MapWise: Accelerated learning through visible thinking*, Network Educational Press

DES (1985), *Mathematics from 5–16*, HMSO

Barrie Galpin and Alan Graham (eds) (2001), *30 Calculator Lessons for Key Stage 3*, A+B Books

Howard Gardner (1979), *Frames of Mind*, Basic Books

Howard Gardner (1993), *Multiple Intelligences: The theory in practice*, Basic Books

Paul Goldenberg *Issues in Mathematics Education*, Education Development Centre website, www2.edc.org

Kath Hart (1980), *Children's Understanding of Mathematics: 11–16*, John Murray

James Hiebert and others (1997), *Making Sense: Teaching and learning mathematics with understanding*, Heinemann

Sue Johnston-Wilder and David Pimm (eds), (2004), *Teaching Secondary Mathematics with ICT*, Open University Press

Candia Morgan (1998), *Writing Mathematically: The discourse of investigation*, Falmer Press

NCET (1998), *Data Capture and Modelling in Mathematics and Science*, NCET

Adrian Oldknow and Ron Taylor (2003), *Teaching Mathematics Using Information and Communications Technology*, Continuum Books

Seymour Papert (1980), *Mindstorms: Children, computers and powerful ideas*, Basic Books

Robert Powell (1997), *Active Whole-Class Teaching*, Robert Powell Publications

Stephanie Prestage and Pat Perks (2001), *Adapting and Extending Secondary Mathematics Activities: New tasks for old*, David Fulton

QCA (2004), 2002/3 *Annual Report on Curriculum and Assessment*, London HMSO

QCA (2004), *Developing Reasoning Through Algebra and Geometry*, DfES

The Royal Society (2001), *The Teaching and Learning of Geometry 11–19*, The Royal Society

Alistair Smith (1998), *Accelerated Learning in Practice*, Network Educational Press

Anne Watson and John Mason (1998), *Questions and Prompts for Mathematical Thinking*, Association of Teachers of Mathematics

Index